Essays about Number Theory
From Classical Topics to New Problems

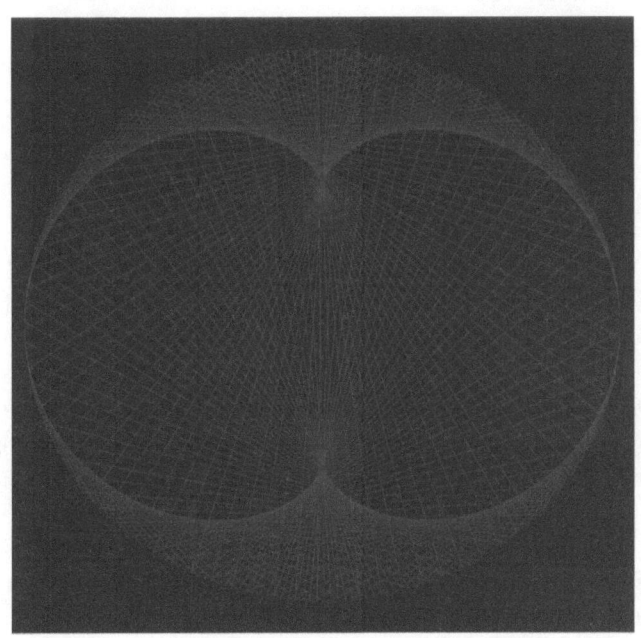

Franz Rothe

Copyright © 2024 by Franz Rothe.

Library of Congress Control Number:		2024900032
ISBN:	Hardcover	979-8-3694-1436-1
	Softcover	979-8-3694-1435-4
	eBook	979-8-3694-1437-8

All rights reserved. No part of this book may be reproduced or transmitted in any form or by any means, electronic or mechanical, including photocopying, recording, or by any information storage and retrieval system, without permission in writing from the copyright owner.

Any people depicted in stock imagery provided by Getty Images are models, and such images are being used for illustrative purposes only. Certain stock imagery © Getty Images.

Print information available on the last page.

Rev. date: 01/02/2024

To order additional copies of this book, contact:
Xlibris
844-714-8691
www.Xlibris.com
Orders@Xlibris.com
802012

Introduction

The *front cover* shows the vertices of a regular 257-gon, connected in the ordering

$$r \mapsto \exp\left[\frac{2\pi i \cdot 3^r}{257}\right]$$

with $r = 0, 1, 2, \ldots 255$. See also formula (7.1.8). Such an ordering of the vertices is used for the construction of the Gaussian periods. On these sums or periods is based the derivation of a closed formula for the coordinates of the vertices, and hence in the end, is proved that the regular 257-gon is constructible with straightedge and compass, in Euclidean geometry.

The *back cover* shows formulas for the x-coordinate of the first vertex in the upper half plane of a regular 17-gon with circum radius 16. The first formula is the one given by Gauss. The second formula I have obtained by my simplified method for the regular 17-gon, which shall be explained on pages 175 to 180. For the simplified method, one does not need to know anything about the Gaussian periods, instead one uses the trigonometric addition theorem and clever guesses.

Preface

This book has been growing over the years out of several resources. Firstly, the courses in number theory and modern algebra I had been teaching at UNC Charlotte. Secondly, my interest to invent mathematical problems for competitions, and individual work with gifted students. It needed some optimism to create this book about number theory. The proofs are gapless and readable, and there are given some exercises with solutions and algorithms. The construction of Fermat polygons is done via computations with mathematica, and a second computer language, finally numeric calculations for check of correctness. Luckily enough, with these means, I could completely work out the constructions of the regular 17, 257 and even the 65 537-gon. This material is much more concrete and computational than any course in number theory or modern algebra usually is.

Complete proof on quadratic reciprocity is given, and the Jacobi symbols are honored in all detail. . Otherwise could be covered just an important classical selection. Generally known problems as well as own problems are included about prime numbers, the Euler totient function and related topics.

MSC2020-Mathematics Subject Classification

11-01 Introductory exposition (textbooks, tutorial papers, etc.) pertaining to number theory

11-04 Software, source code, etc. for problems pertaining to number theory

11A07 Congruences; primitive roots; residue systems

11A15 Power residues, reciprocity

11A25 Arithmetic functions; related numbers; inversion formulas

11A41 Primes

11A51 Factorization; primality

Contents

1 Euclidean Algorithm   **1**
   1.1 The Euclidean Algorithm 1
       1.1.1 The Common Euclidean Algorithm 2
       1.1.2 The Extended Euclidean Algorithm 3
       1.1.3 Further Properties 8
       1.1.4 The Speed of Convergence 11
   1.2 Gauss' Easter Formula 16

2 Primes   **19**
   2.1 The Basics about Primes 19
       2.1.1 Uniqueness of Prime Decomposition 20
       2.1.2 Rational and Irrational 23
       2.1.3 Prime Factorization Program 24
       2.1.4 Sieve of Eratosthenes 25
   2.2 More about Primes 28
       2.2.1 Factoring Factorials 28
       2.2.2 Stirling's Formula 31
       2.2.3 How Many Primes? 35
       2.2.4 Existence of Infinitely Many Primes 37
       2.2.5 Bertrand's Postulate 45
       2.2.6 Infinitely Many Primes under Restrictions . 52
       2.2.7 Infinitely Many Composite Numbers 53
       2.2.8 How Many Owl-Primes Are There? 55
       2.2.9 Arithmetic Progressions of Primes 61

3 Early Achievements   **69**
   3.1 The Chinese Remainder Theorem 69
       3.1.1 Simultaneous Congruences 69
       3.1.2 Chinese Remainder (Sun-Ze's) Theorem . . 74
   3.2 The Geometric Series 75

      3.2.1   My Little Theorems about Powers 75

4 Fermat and Euler   81
4.1 Little Fermat from Iterated Mappings 81
    4.1.1   Moebius Inversion 81
    4.1.2   Using an Iterated Mapping 83
4.2 Euler's Totient Function 86
    4.2.1   The Euler Group 86
    4.2.2   The Euler Totient Function 86
    4.2.3   Euler's Theorem 95

5 On Giants Shoulders   101
    5.0.1   Lagrange's Theorem 103
5.1 Quadratic Residues 104
    5.1.1   Gauss' Proof of Quadratic Reciprocity . . . 107
    5.1.2   Quadratic Residues for Composite Numbers 115
5.2 Jacobi Symbols 119
    5.2.1   Calculation of J-symbols 119
    5.2.2   Jacobi Symbols 123
    5.2.3   Jacobi Symbols in mathematica 125
    5.2.4   Jacobi Symbols and Quadratic Residues . . 129
5.3 Primitive Roots 135
    5.3.1   Primitive Roots, Mainly for Primes 136
    5.3.2   Primitive Roots for Composite Numbers . . 140
    5.3.3   Primitive Roots for some Examples 144
    5.3.4   Semiprimitive Roots for Powers of Two . . . 146
5.4 The Product over the Euler Group 151

6 Around Fermat's Last Problem   155
6.1 The Square Root of a Complex Number 155
6.2 Pythagorean Triples 156
6.3 Towards Fermat's Last Theorem 164

7 Constructible Polygons   169
7.1 Fermat Primes 169
    7.1.1   A Short Paragraph about Fermat Numbers 169
    7.1.2   The 17-gon Construction 173
    7.1.3   Gauss' Polygons 183
    7.1.4   The 257-gon with mathematica 212
    7.1.5   Computations for the 65 537-gon 218
    7.1.6   Identities with Jacobi Symbols 223

8 High Fermat numbers   **229**
    8.0.1   More about Fermat Numbers 229
    8.0.2   Powers of Three 239

9 Prime Testing   **243**
    9.1   Pseudo Primes, Carmichael Numbers 243

Chapter 1

Euclidean Algorithm

1.1 The Euclidean Algorithm

Definition 1 (Division with remainder). Given any two integers a and positive integer $b > 0$, the *quotient* q and *remainder* r are defined as the integers such that

(1.1.1) $$a = q\,b + r \quad \text{and} \quad 0 \leq r < b$$

In that case, we write

(1.1.2) $$a : b = q \quad \text{rem } r$$

and

(1.1.3) $$a \equiv r \mod b$$

Definition 2 (Greatest common divisor). The *greatest common divisor* of two positive integers a and b is the greatest positive integer that is a divisor of both a and b. We denote the greatest common divisor of a and b by $\gcd(a,b)$.

Definition 3 (Least common multiple). The *least common multiple* of two positive integers a and b is the positive integer l such that

(a) l is a multiple of both a and b.

(b) If any other integer k is a multiple of both a and b, then k is a multiple of l.

We denote the least common multiple of a and b by $\operatorname{lcm}(a,b)$.

Intuitively, the greatest common divisor is the *greatest common measure* for the lengths a and b. The least common multiple is the *least common period* of two simultaneous processes with individual periods a and b.

1.1.1 The Common Euclidean Algorithm

Let a, b be positive integers. The greatest common divisor $\gcd(a,b)$ can be calculated by successive divisions with remainders. The algorithm starts with dividing a by b. The last nonzero remainder is the greatest common divisor.

Example 1.1.1. *Take $a = 42, b = 16$.*

$$
\begin{array}{rcll}
42 : & 16 = & 2 \text{ rem} & 10 \\
16 : & 10 = & 1 \text{ rem} & 6 \\
10 : & 6 = & 1 \text{ rem} & 4 \\
6 : & 4 = & 1 \text{ rem} & 2 \\
4 : & 2 = & 2 \text{ rem} & 0
\end{array}
$$

The last nonzero remainder is 2, and hence $\gcd(42, 16) = 2$.

Reason. Let subscript i count the rows. One start with the given numbers as remainders $r_0 = a$ and $r_1 = b$. The i-th row of the scheme is

$$(1.1.4) \qquad r_{i-1} : r_i = q_i \quad \text{rem } r_{i+1}$$

where the successive remainders r_i and quotients q_i satisfy

$$(\text{ri}) \qquad r_{i+1} = r_{i-1} - q_i r_i \quad \text{and} \quad 0 \le r_{i+1} < r_i$$

The algorithm stops when a zero remainder $r_{m+1} = 0$ appears for the first time, say in row m. The last nonzero remainder is r_m. We check that r_m is the greatest common divisor. From the identity

$$(1.1.5) \qquad \gcd(a,b) = \gcd(a - qb, b) = \gcd(b, a - qb)$$

which holds for all integers q, we get inductively

$$\gcd(a,b) = \gcd(r_0, r_1)$$
$$\gcd(a,b) = \gcd(r_{i-1}, r_i) = \gcd(r_i, r_{i+1}) \quad \text{for all } i = 1, 2 \ldots m$$
$$\gcd(a,b) = \gcd(r_m, r_{m+1}) = r_m$$

Hence the last nonzero remainder r_m is the greatest common divisor. $\square$

Problem 1. *Find the greatest common divisor $\gcd(4321, 1234)$.*

1.1.2 The Extended Euclidean Algorithm

Definition 4 (Integer combination). Any number $sa + tb$ with (positive or negative) integers $s, t \in \mathbf{Z}$ is called an *integer combination* of a and b.

Again, a, b are positive integers. Beyond calculating their greatest common divisor, the extended Euclidean algorithm yields the common greatest multiple $\gcd(a, b)$ as an integer combination

(*) $$\gcd(a, b) = sa - tb$$

Actually, the convenient smallest solution s, t of equation (3.1.10) is calculated by the extended Euclidean algorithm. Furthermore, $(-1)^m s \geq 0$ and $(-1)^m t \geq 0$, where m is the number of steps of the algorithm.

Definition 5 (The Extended Euclidean Algorithm). Similar to the common algorithm, one starts with dividing a by b, followed by successive divisions with remainders. One has to keep track of both the quotients q_i and remainders r_i. But the extended algorithm needs two extra columns, which start in row 0 and 1 with the 2×2 unit matrix, whereas the divisions start at row 1 with the calculation of $a : b$. The operation to successively produce these two extra columns is

row two above current row plus quotient × row one above current row $\mapsto$ gives current row.

The algorithm stops when a division has zero remainder for the first time. The greatest common divisor $\gcd(a, b)$ is the last nonzero remainder. In the m-th row, adjacent to the zero remainder r_{m+1}, appear the numbers s_m and t_m such that

(*m) $$(-1)^m \cdot \gcd(a, b) = s_m a - t_m b$$

and hence (3.1.10) follows with $s = (-1)^m s_m$ and $t = (-1)^m t_m$. The optional extra row $m + 1$ does not contain a division, only s_{m+1} and t_{m+1} are calculated. Because of the relations

(1.1.6) $$s_{m+1} = \frac{b}{\gcd(a,b)}, \quad t_{m+1} = \frac{a}{\gcd(a,b)}$$

shown in Corollary 3 below, we get a convenient check.

Example 1.1.2. *Take $a = 42, b = 16$.*

row 0:						1	0
row 1:	42 :	16 =	2 rem	10		0	1
row 2:	16 :	10 =	1 rem	6		1	2
row 3:	10 :	6 =	1 rem	4		1	3
row 4:	6 :	4 =	1 rem	2		2	5
row m=5:	4 :	2 =	2 rem	0		3	8
row m+1:						8	21

Hence $\gcd(42, 16) = 2$ *and* $3 \cdot 42 - 8 \cdot 16 = -2 = -\gcd(42, 16)$. *We get* $s = -3, t = -8$. *These numbers are negative since the number* $m = 5$ *of steps of the algorithm is odd in this example.*

The optional extra row $m+1$ does no longer involve a division step of the ordinary Euclidean algorithm. The extra calculation of s_{m+1} and t_{m+1} is a convenient check. s_{m+1} and t_{m+1} are calculated. Because of the relations

$$(3.1.1) \qquad s_{m+1} = \frac{b}{\gcd(a,b)}, \quad t_{m+1} = \frac{a}{\gcd(a,b)}$$

shown in Corollary 3 below, we get a convenient check.

Reason for the extended algorithm. Let subscript i count the rows. Starting with the given numbers as remainders $r_0 = a$ and $r_1 = b$, the i-th row of the extended scheme is

$$\text{(i-extend)} \qquad r_{i-1} : r_i = q_i \quad \text{rem } r_{i+1} \qquad s_i \quad t_i$$

The two extra columns start with $s_0 = 1, s_1 = 0$ and $t_0 = 0, t_1 = 1$. The extended Euclidean algorithm calculates the successive remainders r_i and quotients q_i via

$$(1.1.7) \qquad r_{i+1} = r_{i-1} - q_i r_i \quad \text{and} \quad 0 \leq r_{i+1} < r_i$$

and uses the recursion to get the sequences s_i and t_i:

$$(1.1.8) \qquad \begin{aligned} s_{i+1} &= s_{i-1} + q_i s_i \\ t_{i+1} &= t_{i-1} + q_i t_i \end{aligned}$$

Since we get the same two step recursion

$$(-1)^{i+1} r_{i+1} = (-1)^{i-1} r_{i-1} + q_i (-1)^i r_i$$

for the alternating remainders $(-1)^i r_i$, we show inductively

$$\text{(combine)} \qquad (-1)^i r_i = s_i a - t_i b$$

1.1. THE EUCLIDEAN ALGORITHM

to hold for all rows $i = 0, 1, \ldots m + 1$. Indeed equation (combine) holds for $i = 0, 1$, and inductively follows for all i, because of formulas (1.1.7) and (1.1.8). The algorithm stops when a zero remainder $r_{m+1} = 0$ appears for the first time in row m. As explained above, the last nonzero remainder is the greatest common divisor.

(1.1.9) $\qquad r_m = \gcd(a, b)\,, \ r_{m+1} = 0$

The identity (combine) with $i = m$ implies

$$(-1)^m r_m = s_m a - t_m b$$

hence formula (*m) holds. The numbers s_m, t_m show up next to the zero remainder $r_{m+1} = 0$. For odd number m steps $s = -s_m, t = -t_m$ are negative, for even m, we get $s = s_m, t = t_m$ which are positive. □

Remark. The above use of the identity (combine) with $i = m$ has been taken from a remark in the article [8].

Problem 2. *Express the greatest common divisor* $\gcd(4321, 1234)$ *as integer combination*

(3.1.10) $\qquad \gcd(a, b) = sa - tb$

of these two given integers.

Answer.

row 0:							1	0
row 1:	4231 :	1234 =	3	rem	619		0	1
row 2:	1234 :	619 =	1	rem	615		1	3
row 3:	619 :	615 =	1	rem	4		1	4
row 4:	615 :	4 =	153	rem	3		2	7
row 5:	4 :	3 =	1	rem	1		307	1 075
row m=6:	3 :	1 =	3	rem	0		309	1 082
row m+1:							1234	4321

Hence $\gcd(4321, 1234) = 1$ and $309 \cdot 4321 - 1\,082 \cdot 1234 = 1 = \gcd(4321, 1234)$. We get $s = 309, t = 1\,082$. These numbers are positive since the number $m = 6$ of steps of the algorithm is even.

Remark (Backtracking "the old fashioned way"). For completeness I mention the backtracking algorithm, which can be used to produce formula (3.1.10), too. At first one does the common Euclidean algorithm, keeping the quotients. But as the price for not planning ahead, one has to remember the two entire sequences of remainders and quotients. Going backwards, one uses the divisions of the algorithm in reversed order and calculates expressions for the greatest common divisor in terms of two successively larger remainders.

Example 1.1.3. *Get integers s, t such that $s42 - t16 = \gcd(42, 16)$.*

$$\begin{aligned}
\gcd(42, 16) = 2 &= \underline{6} - \underline{4} & &= \underline{6} - (\underline{10} - \underline{6}) \\
&= -\underline{10} + 2 \cdot \underline{6} & &= -\underline{10} + 2 \cdot [\underline{16} - \underline{10}] \\
&= 2 \cdot \underline{16} - 3 \cdot \underline{10} & &= 2 \cdot \underline{16} - 3 \cdot [\underline{42} - 2 \cdot \underline{16}] \\
&= -3 \cdot \underline{42} + 8 \cdot \underline{16}
\end{aligned}$$

from which we see one (non unique) solution $s = -3, t = -8$ of formula (3.1.10).

Problem 3. *Use the old fashioned backtracking to express the greatest common divisor*
$\gcd(4321, 1234) = 1$ *as integer combination.*

Answer.

$$\begin{aligned}
\gcd(4321, 1234) &= 1 \\
&= \underline{4} - \underline{3} & &= \underline{4} - [\underline{615} - 153 \cdot \underline{4}] \\
&= -\underline{615} + 154 \cdot \underline{4} & &= -\underline{615} + 154 \cdot [\underline{619} - \underline{615}] \\
&= 154 \cdot \underline{619} - 155 \cdot \underline{615} & &= 154 \cdot \underline{619} - 155 \cdot [\underline{1234} - \underline{619}] \\
&= -155 \cdot \underline{1234} + 309 \cdot \underline{619} & &= -155 \cdot \underline{1234} + 309 \cdot [\underline{4321} - 3 \cdot \underline{1234}] \\
&= 309 \cdot \underline{4321} - 1082 \cdot \underline{1234}
\end{aligned}$$

from which we see one (non unique) solution $s = 309, t = 1082$. such that $s4321 - t1234 = \gcd(4321, 1234)$.

Problem 4. *Calculate the greatest common divisor of 765 and 567. Find integers s and t such that $\gcd(765, 567) = s \cdot 765 - t \cdot 567$.*

Answer. The extended Euclidean algorithms is used to calculate the greatest common divisor. In an additional parallel calculation, one gets the greatest common divisor as integer combination of the

1.1. THE EUCLIDEAN ALGORITHM

two given numbers.

```
row 0:                                    1    0
row 1:      765   567 =   1 rem   198     0    1
row 2:      567 : 198 =   2 rem   171     1    1
row 3:      198 : 171 =   1 rem    27     2    3
row 4:      171 :  27 =   6 rem     9     3    4
row 5:       27 :   9 =   3 rem     0    20   27
row 5+1:                                 63   85
```

Indeed, $\gcd(765, 567) = 9 = (-20) \cdot 765 + 27 \cdot 567$. Hence $s = -20$ and $t = -27$.

Remark. The optional extra row 5+1 does not contain a division, only s_{M+1} and t_{M+1} are calculated. This is a convenient check, since $63 \cdot 765 - 85 \cdot 567 = 0$.

Problem 5. *Use the last problem to calculate the least common multiple* $\operatorname{lcm}(765, 567)$.

Answer.

$$\operatorname{lcm}(765, 567) = 765 \cdot \frac{567}{\gcd(765, 567)} = 765 \cdot 63 = 48\,195$$

Problem 6. *Calculate the greatest common divisor of 367 and 47. Find integers s and t such that* $\gcd(367, 47) = s \cdot 367 + t \cdot 47$.

Answer. The extended Euclidean algorithms is used to calculate the greatest common divisor. In an additional parallel calculation, one gets the greatest common divisor as integer combination of the two given numbers. Here is the example:

```
row 0:                                    1    0
row 1:      367 :  47 =   7 rem    32     0    1
row 2:       47 :  38 =   1 rem     9     1    7
row 3:       38 :   9 =   4 rem     2     1    8
row 4:        9 :   2 =   4 rem     1     5   39
row 5:        2 :   1 =   2 rem     0    21  164
row 5+1:                                 47  367
```

Indeed, $\gcd(367, 74) = 1 = (-21) \cdot 367 + 164 \cdot 47$. The optional extra row $5 + 1$ does not contain a division, only s_{M+1} and t_{M+1} are calculated. This is a convenient check, since $47 \cdot 367 - 367 \cdot 47 = 0$.

1.1.3 Further Properties

The extended Euclidean algorithm is of basic importance, and deserves further observations and Corollaries.

Corollary 1. *The linear diophantine equation*

$$(3.1.10) \qquad \gcd(a,b) = sa - tb$$

has the set of integer solutions s', t'

$$(1.1.10) \qquad s' = s + \lambda \frac{b}{\gcd(a,b)}, \quad t' = t + \lambda \frac{a}{\gcd(a,b)}$$

with arbitrary integer λ.

Reason. It is easy to check that formula (1.1.10) gives integer solutions of equation (3.1.10). Conversely, suppose that both s,t and s',t' are solutions of (3.1.10):

$$sa - tb = \gcd(a,b)$$
$$s'a - t'b = \gcd(a,b)$$

Multiplying the first equation above by t', the second one by t and subtracting yields the first equation below, multiplying the first equation by s', the second one by s and subtracting yields the first equation below. Similarly, we get the second one.

$$(t's - ts')a = (t' - t)\gcd(a,b)$$
$$(-s't + st')b = (s' - s)\gcd(a,b)$$

Hence equation (1.1.10) holds with $\lambda = st' - s't$. □

Corollary 2. *The least common multiple of any positive integers a and b is*

$$(1.1.11) \qquad lcm(a,b) = \frac{a \cdot b}{\gcd(a,b)}$$

Proof. We check that the number

$$(1.1.12) \qquad l := \frac{a \cdot b}{\gcd(a,b)}$$

on the right hand side satisfies both requirements (a) and (b) from the definition 3 of the least common multiple.

1.1. THE EUCLIDEAN ALGORITHM

(a) The number l is a multiple of both a and b, since $l = a \cdot \frac{b}{\gcd(a,b)} = \frac{a}{\gcd(a,b)} \cdot b$.

(b) Assume the positive integer k is a multiple of both a and b. We need to check that k is a multiple of l.

$\square$

Check of item (b). Since the positive integer k is assumed to be a multiple of both a and b, there exist integers p and q such that $k = aq = bp$. The greatest common divisor satisfies

$$(3.1.10) \qquad \gcd(a,b) = sa - tb$$

from the extended Euclidean algorithm. Hence

$$\gcd(a,b) \cdot k = sa \cdot k - tb \cdot k = sa \cdot bp - tb \cdot aq = (a \cdot b) \cdot (sp - tq)$$

$$k = \frac{a \cdot b}{\gcd(a,b)} \cdot (sp - tq) = l \cdot (sp - tq)$$

Hence k is a multiple of l, as to be shown. $\square$

Proposition 1 (The combination and determinant identities).

(combine) $\qquad (-1)^i r_i = s_i a - t_i b \qquad$ for $i = 0, 1 \ldots m+1$
(det) $\qquad s_i t_{i+1} - s_{i+1} t_i = (-1)^i$
(det-r) $\qquad r_i t_{i+1} + r_{i+1} t_i = a \qquad$ for $i = 0, 1 \ldots m$

Problem 7. *Show the determinant identity* (det) *by induction for* $i = 0, 1, \ldots m$. *Use induction step* "$i - 1 \mapsto i$".

Answer. $s_0 t_1 - s_1 t_0 = 1 \cdot 1 - 0 \cdot 0 = 1$ gives the start. Here is the induction step "$i - 1 \mapsto i$":

$$s_i t_{i+1} - s_{i+1} t_i = s_i [t_{i-1} + q_i t_i] - [s_{i-1} + q_i s_i] t_i = s_i t_{i-1} - s_{i-1} t_i$$
$$= -[s_{i-1} t_i - s_i t_{i-1}] = -(-1)^{i-1} = (-1)^i$$

Too, we get formula (det-r), since

$$r_i t_{i+1} + r_{i+1} t_i = (-1)^i [s_i a - t_i b] t_{i+1} + (-1)^{i+1} [s_{i+1} a - t_{i+1} b] t_i$$
$$= (-1)^i a [s_i t_{i+1} - s_{i+1} t_i] = a$$

Corollary 3. *One may take the calculation of the sequences* s_i *and* t_i *one step further up to* $i = m+1$ *and get*

$$((3.1.1)) \qquad s_{m+1} = \frac{b}{\gcd(a,b)}, \qquad t_{m+1} = \frac{a}{\gcd(a,b)}$$

Reason for Corollary 3. We set up a linear system to calculate the extra values s_{m+1} and t_{m+1}. To this end we use formula (combine) for $i = m+1$, and the determinant identity (det) with $i = m$.

(sys) $$\begin{aligned} t_m\, s_{m+1} - s_m\, t_{m+1} &= (-1)^{m+1} \\ a\, s_{m+1} - b\, t_{m+1} &= 0 \end{aligned}$$

The determinant of system (sys) is

$$\Delta = -t_m b + s_m a = (-1)^m \cdot \gcd(a,b) \neq 0$$

Hence system (sys) has a unique solution, to be obtained by Cramer's rule. It turns out to be (3.1.1). □

Remark. Here is a direct calculation to check (3.1.1):

$$\begin{aligned} b &= (-1)^{m+1}[(bs_{m+1}) \cdot t_m - s_m \cdot (bt_{m+1})] \\ &= (-1)^{m+1}[(bs_{m+1}) \cdot t_m - s_m \cdot (as_{m+1})] \\ &= (-1)^{m+1}[bt_m - as_m] \cdot s_{m+1} \\ &= \gcd(a,b) \cdot s_{m+1} \end{aligned}$$

$$\begin{aligned} a &= (-1)^{m+1}[(as_{m+1}) \cdot t_m - s_m \cdot (at_{m+1})] \\ &= (-1)^{m+1}[(bt_{m+1}) \cdot t_m - s_m \cdot (at_{m+1})] \\ &= (-1)^{m+1}[bt_m - as_m] \cdot t_{m+1} = \gcd(a,b) \cdot t_{m+1} \end{aligned}$$

Corollary 4. *Assume $a \neq b$. The solution s,t of equation (3.1.10) constructed by the extended Euclidean algorithm is the <u>unique</u> solution of (3.1.10) satisfying*

(**) $$|s| \leq \frac{b}{2\gcd(a,b)} \quad \text{and} \quad |t| \leq \frac{a}{2\gcd(a,b)}$$

Reason. The assumption $a \neq b$ implies that the last quotient $q_m \geq 2$. Hence corollary 3 and formula (1.1.8) with $i = m$ imply

$$\frac{b}{\gcd(a,b)} = s_{m+1} = s_{m-1} + q_m s_m \geq 2|s|$$

$$\frac{a}{\gcd(a,b)} = t_{m+1} = t_{m-1} + q_m t_m \geq 2|t|$$

for the solution s,t of (3.1.10) constructed above.
Hence s,t satisfy (**). □

1.1. THE EUCLIDEAN ALGORITHM

Problem 8. *Prove uniqueness for the solutions of (3.1.10) and (**).*

Answer. Suppose that s, t and s', t' are different solutions of formulas (3.1.10) and (**).
Both

$$|s' - s| = |\lambda| \frac{b}{\gcd(a,b)} \geq \frac{b}{\gcd(a,b)} \quad \text{and}$$

$$|s' - s| \leq |s| + |s'| \leq 2 \frac{b}{2\gcd(a,b)}$$

by (1.1.10) from Corollary 1, and formula (**). Since $s \neq s'$, both inequalities together imply $|s' - s| = |s| + |s'|$ and hence $s = -s' = \pm \frac{b}{2\gcd(a,b)}$. Similarly one gets $t = -t' = \pm \frac{a}{2\gcd(a,b)}$. Now formula (3.1.10) implies

$$\gcd(a,b) = sa - tb = -(s'a - t'b) = -\gcd(a,b) = 0$$

which is impossible.

1.1.4 The Speed of Convergence

We now investigate how fast the Euclidean algorithm is. One needs the Fibonacci sequence

$$F_0 = 0, \ F_1 = 1, \ F_2 = 1, \ F_3 = 2, \ F_4 = 3, \ F_5 = 5, \ F_6 = 8, \ldots$$
$$F_{i+1} = F_i + F_{i-1} \quad \text{for all } i \geq 1$$

Proposition 2. *Let $q \geq 0, m \geq 1$ be any integers. The Euclidean algorithm needs exactly m steps for the following examples:*

(1) $a = F_{m+2} + qF_{m+1}$, $b = F_{m+1}$, $m \geq 1$
(2) $a = F_m$, $b = F_{m+1} + qF_m$, $m \geq 2$

Problem 9. *Check these statements, extended mode, with*

(a) $m = 6, q = 2$ *and*

(b) $m = 6, q = 0$.

CHAPTER 1. EUCLIDEAN ALGORITHM

Answer. (a) $m = 6$, $q = 2$

(1.1.13)

row 0:						1	0
row 1:	47 :	13 =	3 rem	8		0	1
row 2:	13 :	8 =	1 rem	5		1	3
row 3:	8 :	5 =	1 rem	3		1	4
row 4:	5 :	3 =	1 rem	2		2	7
row 5:	3 :	2 =	1 rem	1		3	11
row m=6:	2 :	1 =	2 rem	0		5	18
row m+1:						13	47

(1.1.14)

row 0:						1	0
row 1:	8 :	29 =	0 rem	8		0	1
row 2:	29 :	8 =	3 rem	5		1	0
row 3:	8 :	5 =	1 rem	3		3	1
row 4:	5 :	3 =	1 rem	2		4	1
row 5:	3 :	2 =	1 rem	1		7	2
row m=6:	2 :	1 =	2 rem	0		11	3
row m+1:						29	8

(b) $m = 6$, $q = 0$

(1.1.15)

row 0:						1	0
row 1:	21 :	13 =	1 rem	8		0	1
row 2:	13 :	8 =	1 rem	5		1	1
row 3:	8 :	5 =	1 rem	3		1	2
row 4:	5 :	3 =	1 rem	2		2	3
row 5:	3 :	2 =	1 rem	1		3	5
row m=6:	2 :	1 =	2 rem	0		5	8
row m+1:						13	21

(1.1.16)

row 0:						1	0
row 1:	8 :	13 =	0 rem	8		0	1
row 2:	13 :	8 =	1 rem	5		1	0
row 3:	8 :	5 =	1 rem	3		1	1
row 4:	5 :	3 =	1 rem	2		2	1
row 5:	3 :	2 =	1 rem	1		3	2
row m=6:	2 :	1 =	2 rem	0		5	3
row m+1:						13	8

Reason for proposition 2. For example (1.1.15), the Euclidean al-

1.1. THE EUCLIDEAN ALGORITHM

gorithm consists of these m steps:
(1.1.17)

row 1:	a :	$b =$	$q + 1$	rem	F_m
row 2:	b :	$F_m =$	1	rem	F_{m-1}
row 3:	F_m :	$F_{m-1} =$	1	rem	F_{m-2}
row i:	F_{m+3-i} :	$F_{m+2-i} =$	1	rem	F_{m+1-i}
row m-1:	F_4 :	$F_3 =$	1	rem	1
row m:	F_3 :	$F_2 =$	2	rem	0

For example (1.1.16), only the two first steps are different. Here are all m steps:
(1.1.18)

row 1:	a :	$b =$	0	rem	F_m
row 2:	b :	$F_m =$	$q + 1$	rem	F_{m-1}
row 3:	F_m :	$F_{m-1} =$	1	rem	F_{m-2}
row i:	F_{m+3-i} :	$F_{m+2-i} =$	1	rem	F_{m+1-i}
row m-1:	F_4 :	$F_3 =$	1	rem	1
row m:	F_3 :	$F_2 =$	2	rem	0

$\square$

Proposition 3 (Convergence speed of the Euclidean algorithm). *Let a and b be positive integers. If the Euclidean algorithm needs m steps, then*

(1.1.19) $\qquad b \geq \gcd(a,b)\, F_{m+1}$

(1.1.20) $\qquad a \geq \gcd(a,b)\, F_{m+2} \quad \text{if } a > b$

Proof. The claim is obvious for $m = 1$. In this case, b divides a and $b = \gcd(a,b)$. Suppose the Euclidean algorithm needs $m \geq 2$ steps. Hence the last quotient $q_m \geq 2$. Following the remainders backwards yields

$$r_m = \gcd(a,b) = \gcd(a,b)\, F_2$$
$$r_{m-1} \geq 2r_m = \gcd(a,b)\, F_3$$
$$r_{m-2} \geq r_m + r_{m-1} \geq \gcd(a,b)\, F_4$$

and inductively for $i = m-1, m-2 \ldots 2$

(rup) $\qquad r_{i-1} \geq r_{i+1} + q_i r_i \geq r_{i+1} + r_i \geq \gcd(a,b)\, F_{m+3-i}$

For $i = 2$, we get the first claim $b \geq \gcd(a,b)\, F_{m+1}$. If $a > b$, we know that $q_1 \geq 1$, and hence $i = 1$ can be included in the estimate (rup). Thus one confirms the second claim $a \geq \gcd(a,b)\, F_{m+2}$. $\square$

Proposition 4 (Convergence speed of the Euclidean algorithm). *Let a and b be positive integers. If*

(Fn)
  either $\quad b \leq F_{n+1}$ with $n \geq 1 \quad$ or $\quad a \leq F_n$ with $n \geq 2$

the number m of steps in the Euclidean algorithm satisfies $m \leq n$. Under assumption (Fn), *the algorithm takes $m = n$ steps exactly for the examples (1) and (2) given in proposition 2 above.*

Proof. If $b \leq F_2 = 1$ and $n = 1$, then $m = 1$ is obvious. Now suppose that $b \leq F_{n+1}$ and $m \geq n \geq 2$. From estimate (1.1.19), we conclude

(1.1.21) $\qquad F_{m+1} \leq \gcd(a,b)$, $F_{m+1} \leq b \leq F_{n+1} \leq F_{m+1}$

Hence $m = n$, $b = F_{m+1}$, and $\gcd(a,b) = 1$. Equality holds for the entire sequence of estimates (rup) and hence $q_i = 1$ for $i = 2 \ldots m-1$ and $q_m = 2$. One gets $r_2 = \gcd(a,b) F_m = F_m$, hence $a = q_1 b + r_2 = q_1 F_{m+1} + F_m = (q_1 - 1) F_{m+1} + F_{m+2}$. In case that $q_1 \geq 1$, we get example (1). Or, as an exceptional case $q_1 = 0$, we conclude $a = r_2 = F_m$ and $b = F_{m+1}$, which is example (2) with $q = 0$.

If $a \leq F_2 = 1$ and $n = 2$, then $m \leq 2$ is obvious. Secondly, suppose that $a \leq F_n$ and $m \geq n \geq 3$. Distinguish the cases (a):$a \geq b$ and (b):$a < b$. Case (a) cannot occur, because $\gcd(a,b) F_{m+2} \leq a \leq F_n \leq F_m$ leads to a contradiction.

In case (b), the first quotient is zero and first three remainders are $r_0 = a, r_1 = b, r_2 = a$. After discarding the first step, we get a Euclidean algorithm with $m - 1 \geq n - 1 \geq 2$ steps, for which the first entry is now b and second entry is $a \leq F_n$. As explained above, one concludes

(1.1.22) $\qquad F_m \leq \gcd(b,a) F_m \leq a \leq F_n \leq F_m$

Hence $m = n$ and $a = F_m$, $\gcd(a,b) = 1$, $r_3 = F_{m-1}$. Because of $b = r_1 = q_2 r_2 + r_3 = (q+1) r_2 + r_3 = (q+1) F_m + F_{m-1} = q F_m + F_{m+1}$, we get example (2). $\square$

The *golden number* ϕ and the explicit formula for the Fibonacci numbers are

(1.1.23) $\qquad \phi := \dfrac{\sqrt{5}+1}{2}, \qquad F_n = \dfrac{\phi^n - (-1)^n \phi^{-n}}{\sqrt{5}}$

This formula can be inverted. For all $n \geq 2$, one gets

(Fibinverse) $\qquad n = \left\lceil \dfrac{\log F_n - \log 2}{\log \phi} \right\rceil + 3$

1.1. THE EUCLIDEAN ALGORITHM

Problem 10. *Calculate the Fibonacci numbers and check the inversion formula* (Fibinverse) *for* $n \leq 12$.

Theorem 1 (Logarithmic effectiveness of the Euclidean Algorithm). *The number of steps $m(a,b)$ the Euclidean algorithm takes is bounded above by*

(1.1.24)
$$m(a,b) \leq \min\left(\left\lceil \frac{\log a - \log 2 - \log \gcd(a,b)}{\log \phi} + 3 \right\rceil, \left\lceil \frac{\log b - \log 2 - \log \gcd(a,b)}{\log \phi} + 2 \right\rceil\right)$$

Independent verification. We may assume $a \neq b$ since the case $a = b$ is obvious. Because of $q_i \geq 1$ for $2 \leq i \leq m$ and even $q_m \geq 2$, equation (1.1.8) implies inductively $s_i \geq F_{i-1}$ for $i = 1, 2 \ldots m$. Corollary 3 implies now

(1.1.25)
$$\frac{b}{\gcd(a,b)} = s_{m+1} \geq 2s_m + s_{m-1} \geq 2F_{m-1} + F_{m-2} = F_{m+1}$$

with $m = m(a,b)$. Hence the inversion formula (Fibinverse) implies

(1.1.26)
$$m(a,b) + 1 = \left\lceil \frac{\log F_{m+1} - \log 2}{\log \phi} \right\rceil + 3$$
$$\leq \left\lceil \frac{\log b - \log 2 - \log \gcd(a,b)}{\log \phi} \right\rceil + 3$$

To get an estimate of $m(a,b)$ in terms of a, we distinguish the cases

(a): $a < b$

(b): $a > b$

(c): $a = b$

In case (a), the algorithm starts with the remainders a, b, a. Discarting the first step yields

(1.1.27)
$$m(a,b) - 1 = m(b,a)$$
$$\leq \left\lceil \frac{\log a - \log 2 - \log \gcd(a,b)}{\log \phi} + 2 \right\rceil$$

as claimed. In case (b), we consider the prolonged algorithm starting with the remainders $a+b, a, b$ from which one gets

(1.1.28)
$$m(a,b) = m(a+b, a) - 1$$
$$\leq \left\lceil \frac{\log a - \log 2 - \log \gcd(a,b)}{\log \phi} + 1 \right\rceil$$

which yields an estimate of $m(a,b)$ better—by two—than the one claimed. □

1.2 Gauss' Easter Formula

Definition 6 (Greatest integer or floor function). For any real x, the floor of x is the unique integer such that $n \leq x < n+1$. The floor of x is denoted by $\lfloor x \rfloor$.

Definition 7 (ceiling function). For any real x, the ceiling of x is the unique integer such that $n-1 < x \leq n$. The ceiling of x is denoted by $\lceil x \rceil$.

The exact calculation of the date for the easter festival has always been of great importance for the church. In the Middle Ages there existed specialists called "computists" with the task to calculate the date of easter, or "computus paschalis". This was a complex endeavor involving many tables and fine points of the calender.

In the year 1800, the German mathematician Carl Friedrich Gauß published a simplified method for calculating the date of easter, but his method still looks quiet involved:

$$M = 24, N = 5$$
$$a = J \pmod{4}$$
$$b = J \pmod{7}$$
$$c = J \pmod{19}$$
$$d = (19c + M) \pmod{30}$$
$$e = (2a + 4b + 6d + N) \pmod{7}$$
$$f = \lfloor \frac{c + 11d + 22e}{451} \rfloor$$
$$Ostern = 22 + d + e - 7f$$

Here J means the year for which the date of easter is to be calculated. By $\pmod{m}$ is denoted the modular function, being the remainder after division by the number m. For example 2019 $\pmod 4$ = 3. Hence for the actual year $J = 2019$ one gets b = 2019 $\pmod 7$ = 3 and c = 2019 $\pmod{19}$ = 5. For the calculation of the numbers d and e, we need the constants M and N which depend on the calender. For this century, Gauss gives the values $M = 25$ and $N = 5$. For he example we obtain $d = (19c + M) \pmod{30} = 29$

1.2. GAUSS' EASTER FORMULA

and $e = (2a + 4b + 6d + N) \pmod 7 = 1$. The computation of f involves the floor function, defined above by definition 6. For the example one obtains

$$f = \left\lfloor \frac{c + 11d + 22e}{451} \right\rfloor = \left\lfloor \frac{5 + 11 \cdot 29 + 22 \cdot 1}{451} \right\rfloor$$

$$= \left\lfloor \frac{346}{451} \right\rfloor = 0$$

$22 + d + e - 7f = 52$

Since this is not a date in march, easter will be in april, indeed on the $52 - 31 = 21$-th april. Included in this small section is a program in mathematica and a table.

year	date
2000	april, 23
2001	april, 15
2002	march, 31
2003	april, 20
2004	april, 11
2005	march, 27
2006	april, 16
2007	april, 8
2008	march, 23
2009	april, 12
2010	april, 4
2011	april, 24
2012	april, 8
2013	march, 31
2014	april, 20

year	date
2015	april, 5
2016	march, 27
2017	april, 16
2018	april, 1
2019	april, 21
2020	april, 12
2021	april, 4
2022	april, 17
2023	april, 9
2024	march, 31
2025	april, 20
2026	april, 5
2027	march, 28
2028	april, 16
2029	april, 1

year	date
2030	april, 21
2031	april, 13
2032	march, 28
2033	april, 17
2034	april, 9
2035	march, 25
2036	april, 13
2037	april, 5
2038	april, 25
2039	april, 10
2040	april, 1
2041	april, 21
2042	april, 6
2043	march, 29
2044	april, 17
2045	april, 9
2046	march, 25
2047	april, 14
2048	april, 5
2049	april, 18
2050	april, 10

```
Ostern = Function[J,
MM=24;NN=5;
a=Mod[J,4];b=Mod[J,7]; c=Mod[J,19];
d=Mod[(19c+MM),30];
e=Mod[(2a+4b+6d+NN),7];
f= Floor[(c+11d+22e)/451];
22+d+e-7f];

mar= Function[J,Ost = Ostern[J];If[Ost<=31, Ost,Null]];
apr=Function[J,Ost=Ostern[J];If[Ost>31,Ost-31,Null]];
```

Chapter 2

Primes

2.1 The Basics about Primes

Proposition 5 (Euclidean Property). *If a number c divides the product ab and $\gcd(c,a) = 1$, then c divides b.*

Standard proof. By the extended Euclidean algorithm, there exist integers s, t such that

$$1 = sa + tc$$
$$\frac{b}{c} = s\frac{ab}{c} + tb$$

The second line results by multiplication of both sides with $\frac{b}{c}$. Because c divides ab, the right hand side is an integer. Hence c divides b, as to be shown. $\square$

Second proof. Both numbers a and c are divisors of both products ab and ac. Hence both a and c are divisors of the greatest common divisor

$$G := \gcd(ab, ac)$$

Hence the integer

$$(2.1.1) \qquad q := \frac{ac}{G}$$

is a divisor of both a and c. Hence q is a divisor of $\gcd(a,c) = 1$, which was assumed to be one. Hence $q = 1$, and $ac = G$ is a divisor of ab. This implies that c is a divisor of b, as to be shown. $\square$

Definition 8 (prime number). A *prime number* is an integer $p \geq 2$, which is divisible only by 1 and itself.

Theorem 2 (Euclid). *There exist infinitely many primes.*

Proof. Put the first k primes into the increasing sequence p_i and let
$$P = \prod_{1 \leq i \leq k} p_i$$
The number $P+1$ may be a prime or composite. In the first case, we have found a prime larger than p_k. In he second case, the number $P+1$ may be prime factored. All its prime factors are larger that p_k. In both cases, we have shown that there exist at least $k+1$ primes. Since this argument holds for all natural numbers k, the number of primes is not finite. □

Euclid—and many other mathematicians—have shown there exist infinitely many prime numbers. We put them into the increasing sequence
$$p_1 = 2,\ p_2 = 3,\ p_3 = 5,\ p_4 = 7, \ldots$$
It is rather easy to see that for every positive integer, there exists a decomposition into prime factors. Let a and b be any positive integers. There exist sequence $\alpha_i \geq 0$ and $\beta_i \geq 0$, with index $i = 1, 2, \ldots$ and only finitely many terms nonzero such that

(2.1.2) $$a = \prod_{i \geq 1} p_i^{\alpha_i},\quad b = \prod_{i \geq 1} p_i^{\beta_i}$$

2.1.1 Uniqueness of Prime Decomposition

The uniqueness of the prime decomposition turns out harder to prove. Astonishingly, the proof depends on Euclid's lemma, the proof of which in turn relies on the extended Euclidean algorithm.

Proposition 6 (Euclid's Lemma). *If a prime number divides the product of two integers, the prime number divides at least one of the two integers.*

Reason. Let p be the prime number, and the integers be a and b. We assume that p divides the product ab, but p does not divide a. We need to show that p divides b.

2.1. THE BASICS ABOUT PRIMES

Because p does not divide a, the definition of a prime number implies $\gcd(a,p) = 1$. By the extended Euclidean algorithm, there exist integers s, t such that

$$1 = sa + tp$$

Hence

$$\frac{b}{p} = s\frac{ab}{p} + tb$$

Because p divides ab, the right hand side is an integer. Hence p divides b, as to be shown. □

Proposition 7 (Monotonicity). *Let a and b have the prime decompositions*

(2.1.3) $$a = \prod_{i \geq 1} p_i^{\alpha_i} \, , \quad b = \prod_{i \geq 1} p_i^{\beta_i}$$

The number b is a divisor of a if and only if $\beta_i \leq \alpha_i$ for all $i \geq 1$.

Reason. If $\beta_i \leq \alpha_i$ for all $i \geq 1$, then $a = qb$ with

$$q = \prod_{i \geq 1} p_i^{\alpha_i - \beta_i}$$

and hence b is a divisor of a.

Conversely, assume that b is a divisor of a. We need to show that $\beta_i \leq \alpha_i$ for all $i \geq 1$. Proceed by induction on b. If $b = 1$, then $\beta_i = 0$ for all $i \geq 1$, and the assertion is true.

Here is the induction step "$b < n \mapsto b = n$": Let p_i be any prime factor of $b \geq 2$, which means that $\beta_i \geq 1$. Because p_i divides b and b divides a, the prime p_i divides a. By Euclid's Lemma Proposition 6, p_i is a divisor of one of the primes p_j occurring in the prime decomposition of a. Hence $\alpha_j \geq 1$. Because different primes cannot divide each other, this implies $i = j$ and $p_i = p_j$. Hence $\frac{b}{p_j} < n$ is a divisor of $\frac{a}{p_j}$. By the induction assumption, this implies $\beta_i \leq \alpha_i$ for all $i \neq j$, as well as $\beta_j - 1 \leq \alpha_j - 1$ and hence $\beta_i \leq \alpha_i$ for all $i \geq 1$. □

Proposition 8 (Uniqueness of prime decomposition). *The prime decomposition of any positive integer is unique.*

Reason. Assume

$$a = \prod_{i \geq 1} p_i^{\alpha_i} \quad \text{and} \quad a = \prod_{i \geq 1} p_i^{\beta_i}$$

Because a divides a, the fact given above both tells that $\beta_i \leq \alpha_i$ and $\alpha_i \leq \beta_i$ for all $i \geq 1$. Hence $\beta_i = \alpha_i$ for all $i \geq 1$. □

Proposition 9. *Let a and b have the prime decompositions (2.1.3). The prime decompositions of the greatest common divisor and least common multiple are*

$$\text{(2.1.4)} \qquad \gcd(a,b) = \prod_{i \geq 1} p_i^{\min[\alpha_i, \beta_i]}$$

$$\text{(2.1.5)} \qquad \text{lcm}\,(a,b) = \prod_{i \geq 1} p_i^{\max[\alpha_i, \beta_i]}$$

Problem 11. *Check these formulas for $a = 1001, b = 4221$.*

Proof of Proposition 9. Let g be the righthand side of equation (2.1.5). We need to check properties (i) and (ii) defining the greatest common divisor.

(i) g divides both a and b.

> *Check.* This is clear, because both $\min[\alpha_i, \beta_i] \leq \alpha_i$ and $\min[\alpha_i, \beta_i] \leq \beta_i$ for all $i \geq 1$. □

(ii) If any positive integer h divides both a and b, then h divides the greatest common divisor g.

> *Check.* Let
>
> $$\text{(2.1.6)} \qquad h = \prod_{i \geq 1} p_i^{\gamma_i}$$
>
> be the prime decomposition of h. Because h divides both a and b, monotonicity implies that both $\gamma_i \leq \alpha_i$ and $\gamma_i \leq \beta_i$ for all $i \geq 1$. Hence $\gamma_i \leq \min[\alpha_i, \beta_i]$ for all $i \geq 1$, which easily implies that h is a divisor of g. □

Let l be the righthand side of equation (2.1.5). We need to check properties (i) and (ii) defining the least common multiple.

(i) The number l is a multiple of both a and b.

> *Check.* This is clear, because both $\max[\alpha_i, \beta_i] \geq \alpha_i$ and $\max[\alpha_i, \beta_i] \geq \beta_i$ for all $i \geq 1$. □

2.1. THE BASICS ABOUT PRIMES

(ii) If any positive integer k is a multiple of both a and b, the integer k is a multiple of the least common multiple l.

Check. Let

$$(2.1.7) \qquad k = \prod_{i \geq 1} p_i^{\gamma_i}$$

be the prime decomposition of k. Because k is a multiple of both a and b, monotonicity implies that both $\gamma_i \geq \alpha_i$ and $\gamma_i \geq \beta_i$ for all $i \geq 1$. Hence $\gamma_i \geq \max[\alpha_i, \beta_i]$ for all $i \geq 1$, which easily implies that k is a multiple of l. □

□

2.1.2 Rational and Irrational

Proposition 10. *For any natural numbers $r \geq 1$ and $a \geq 1$, the r-th root $\sqrt[r]{a}$ is only a rational number if it is even an integer.*

Proof. The assertion is clear for $r = 1$ or $a = 1$, hence we may assume $r \geq 2$ and $a \geq 2$. Now assume that

$$\sqrt[r]{a} = \frac{m}{n}$$
$$n^r a = m^r$$

with natural m, n. We show that any prime number p dividing n has to divide m, too. Hence after cancelling common factors, we get $n = 1$, and hence the root is an integer.

Now assume that the prime p divides n. Hence p^r divides n^r, which in turn divides $an^r = m^r$. Hence p^r divides m^r. Hence, by Euclid's Lemma p divides m, as claimed. As already explained, the assertion follows. □

Proposition 11. *For any natural numbers $r \geq 2$ and any $a \geq 2$, which is not the r-th power of an integer, the root $\sqrt[r]{a}$ is irrational. Especially, the roots of the primes are all irrational.*

Proof. If the root $\sqrt[r]{a}$ is rational, it is even an integer m, and hence $a = m^r$ is the r-th power of m.

Take the contrapositive: If a is not the r-th power of an integer, the root $\sqrt[r]{a}$ is irrational. □

2.1.3 Prime Factorization Program

(Prime factorization program for the TI84). *The program asks for the number N and outputs its prime factorization.*

Unfortunately, in the version written down below, I needed to replace list L_1 by L1, and $\neq$ by not= . Please find out from the context.

PROGRAM:FAK

```
Prompt N
abs(N) -> X
{N/X} -> L1
gcd(X,2) -> P
If P=1
Goto 9
Lbl 0
int(X/P) -> Q
If  P*Q-X not= 0
Goto 9
augment(L1,{P}) -> L1
Q -> X
Goto 0
Lbl 9
2+P-(P=2) -> P
If P^2 <= X
Goto 0
If X not= 1
augment(L1,{X}) -> L1
L1
```

I have tried to produce a program with the following requirements:

- *The program gives an error message for input 0*

- *The program gives an error message for input a too large number to be handled, or not an integer*

- *The program does not run into an endless loop for any input*

- *For a positive integer, the program puts $\{1, prim factors\}$ into the list L_1*

- *For a negative integer, the program puts $\{-1, primfactors\}$ into the list L_1*
- *The output list can be strolled at output*
- *The program gives the prime factorization in at most 15 minutes*

Problem 12. *Calculate and prime factor the number $2^{15} - 1 - (2^3 - 1)(2^5 - 1)$.*

Answer. $(2^p - 1)$ is always a divisor of $(2^{p \cdot q} - 1$ for any p and q. The prime factoring is

$$2^{15} - 1 - (2^3 - 1)(2^5 - 1) = 7 \cdot 31 \cdot \left[\frac{32\,767}{7 \cdot 31} - 1\right] = 7 \cdot 31 \cdot 150$$
$$= 2 \cdot 3 \cdot 5^2 \cdot 7 \cdot 31 = 32\,550$$

2.1.4 Sieve of Eratosthenes

Let N be a (big) natural number. For the sieve of Eratosthenes, we list the numbers $1, 2, \ldots, N$. We successively scratch through (delete) all proper multiples of primes $q \leq \sqrt{N}$. The remaining numbers are 1, and the primes $\sqrt{N} < p \leq N$.

How many numbers did we delete in the sieving? Legendre used the inclusion-exclusion principle for a recount, and in 1808 came up with Proposition 12 below. We need the Moebius function from definition 13 and define

$$\mathcal{P} = \{\, q : q \leq \sqrt{N} \text{ is a prime}\,\}$$
$$\mathcal{D} = \{\, \prod q : q \leq \sqrt{N} \text{ are all different primes}\,\}$$

$1 \in \mathcal{D}$ for the empty product. Let $\pi(n)$ be the number of primes less or equal to n.

Proposition 12 (Legendre's counting formula).

(2.1.8) $\qquad \pi(N) = \pi(\sqrt{N}) - 1 + \{\, \sum \mu(d) \lfloor \frac{N}{d} \rfloor : 1 \leq d \in \mathcal{D}\,\}$

gives the number of primes up to N.

Corollary 5. *Let $f(n)$ be any function defined on the integers and define the sums*

$$F(n) := \{\, \sum f(p) : p \leq n \text{ is a prime}\,\}$$

These sums satisfy

(2.1.9)
$$F(N) = F(\sqrt{N}) - f(1) + \{\sum \mu(d) \left(\sum_{1 \leq jd \leq N} F(jd)\right) : 1 \leq d \in \mathcal{D}\}$$

Especially, we get for Chebyshev sum of logarithms of primes

(2.1.10)
$$\theta(N) = \{\sum \log q : q \leq N \text{ is a prime}\}$$

(2.1.11)
$$\theta(N) = \theta(\sqrt{N}) + \{\sum \mu(d) \left(\lfloor \tfrac{N}{d} \rfloor \log d + \log \lfloor \tfrac{N}{d} \rfloor !\right) : 1 \leq d \in \mathcal{D}\}$$

Lemma 1 (Inclusion-exclusion principle). *For any finite sets $A, B, C, D, \ldots$*

$$|A \cup B| = |A| + |B| - |A \cap B|$$
$$|A \cup B \cup C| = |A| + |B| + |C| - |A \cap B| - |A \cap C| - |B \cap C|$$
$$+ |A \cap B \cap C|$$

$$|A \cup B \cup C \cup D| = |A| + |B| + |C| + |D|$$
$$-|A \cap B| - |A \cap C| - |A \cap D| - |B \cap C| - |B \cap D| - |C \cap D|$$
$$+|A \cap B \cap C| + |A \cap B \cap D| + |A \cap C \cap D| + |B \cap C \cap D|$$
$$-|A \cap B \cap C \cap D|$$

and so on for any finite union.

Problem 13. *Use the inclusion-exclusion principle to count the numbers less or equal to 100 which are divisible either by 2 or 3.*

Solution. Let

$$A := \{n : 1 \leq n \leq 100 \text{ , and 2 divides } n\} \text{ and}$$
$$B := \{n : 1 \leq n \leq 100 \text{ , and 3 divides } n\}$$

Hence $|A \cup B| = |A| + |B| - |A \cap B| = \lfloor \tfrac{100}{2} \rfloor + \lfloor \tfrac{100}{3} \rfloor - \lfloor \tfrac{100}{6} \rfloor = 50 + 33 - 16 = 67$. □

Legendre's counting formula. How many numbers did we delete while doing the sieving of Eratosthenes? Indeed, we did delete

2.1. THE BASICS ABOUT PRIMES

all primes $2 \le q \le \sqrt{N}$, and all composite numbers $1 < n \le N$. Hence

$$\pi(\sqrt{N}) + N - 1 - \pi(N) \quad \text{counts the numbers deleted}$$

On the other hand, we use the inclusion-exclusion principle to count the deleted numbers. We see that

$$\{\sum (-\mu(d))\lfloor \tfrac{N}{d} \rfloor \ : \ 1 < d \in \mathcal{D}\} \quad \text{counts the numbers deleted}$$

Since these two numbers are equal, we get Legendre's formula. □

Problem 14. *Use Legendre's formula to get the number of primes less equal $p \le 100$.*

products	$\lfloor \tfrac{N}{d} \rfloor$	$\sum \mu(d) \lfloor \tfrac{N}{d} \rfloor$
1	100	+100
2	50	−117
3	33	
5	20	
7	14	
2·3	16	+45
2·5	10	
2·7	7	
3·5	6	
3·7	4	
5·7	2	
2·3·5	3	−6
2·3·7	2	
2·5·7	1	
3·5·7	0	
$\pi(\sqrt{N})$ −1		4 −1

There are $\pi(100) = \pi(10) - 1 + 100 - 117 + 45 - 6 = 25$ primes up to 100.

2.2 More about Primes

2.2.1 Factoring Factorials

Lemma 2. *Any number $n \geq 2$ is a prime if and only if*
$$n \mid \binom{n}{k} \quad \text{for } 1 \leq k \leq n-1$$
For any composite number $n \geq 4$ and any proper prime divisor $q \mid n$, $1 < q < n$ holds $n \nmid \binom{n}{q}$.

Proof. Assume $n = p$ is a prime and let $1 \leq k \leq n-1$. The binomial coefficient
$$\binom{p}{k} = \frac{p(p-1)\cdots(p-k+1)}{1 \cdot 2 \cdots k}$$
has the numerator divisible by p, but the denominator not divisible by p. Hence $p \mid \binom{p}{k}$.

Conversely, we assume statement (2) to hold. Let $q \mid n$ be any proper prime divisor of n. Since $q < n$ we get from the assumption
$$n \mid \binom{n}{q} = \frac{n}{q} \cdot \frac{(n-1)\cdots(n-q+1)}{1 \cdot 2 \cdots (q-1)} \quad \text{and hence}$$
$$q \mid \frac{(n-1)\cdots(n-q+1)}{1 \cdot 2 \cdots (q-1)} = \binom{n-1}{q-1}$$
In the left-hand fraction no factor in the denominator is divisible by the prime q. Too, no factor in the numerator is divisible by q since q is a prime and $q \mid n$. Hence we obtain a contradiction. The only way out is $q = n$, in which case no divisibility is assumed to hold, but indeed $n \nmid \binom{n}{n} = 1$. Hence n has no proper prime divisors and hence is a prime. □

Proposition 13 (Legendre). *Let $N \geq 1$ be any natural number and p be any prime. The highest prime power dividing the factorial is*

(2.2.1) $$p^r \| N! \quad \text{with } r = \sum_{l \geq 1} \lfloor \frac{N}{p^l} \rfloor$$

Proof. The factorial $N!$ contains the $\lfloor \frac{N}{p} \rfloor$ factors $1 \cdot p, 2 \cdot p, \ldots, \lfloor \frac{N}{p} \rfloor \cdot p$ divisible by prime p. The $\lfloor \frac{N}{p^2} \rfloor$ factors $1 \cdot p^2, 2 \cdot p, \ldots, \lfloor \frac{N}{p} \rfloor \cdot p^2$ divisible by prime p^2 give $\lfloor \frac{N}{p^2} \rfloor$ extra factors p.

Taking into account all prime powers p^l results in the sum in formula (2.2.1). □

2.2. MORE ABOUT PRIMES

Notation. $p^r \| a$ means that p^r is the *highest* prime power of p dividing the integer a.

Proposition 14 (Lucas). *Let $n = a+b \geq 1$ be any natural number and p be any prime. We write a and b in p-adic representations. The highest prime power dividing the binomial coefficient is*

$$(2.2.2) \qquad p^t \| \binom{n}{a}$$

where t is the number of carry-overs needed for the addition $n = a + b$ done in p-adic representation.

Proof. Let the p-adic representations of a be

$$a = \sum_{s \geq 0} a_s p^s \text{ with } 0 \leq a_s < p \text{ for all } s \geq 0$$

and for b and n accordingly. Let $c_0 = 0$ and $c_s \in \{0,1\}$ be the carry-overs for the p-adic addition $n = a + b$. According to the algorithm for the p-adic addition holds for $s \geq 0$

$$m_s = a_s + b_s + c_s$$

$$n_s = \begin{cases} m_s & \text{if } m_s < p; \\ m_s - p & \text{if } m_s \geq p \end{cases} \quad \text{and} \quad c_{s+1} = \begin{cases} 0 & \text{if } m_s < p; \\ 1 & \text{if } m_s \geq p \end{cases}$$

$$n_s = a_s + b_s + c_s - p c_{s+1}$$

The prime powers in the factorials are according to formula (2.2.1) from
Legendre's proposition 13

$$p^\alpha \| a! \text{ with } \alpha = \sum_{l \geq 1} \lfloor \frac{a}{p^l} \rfloor = \sum_{l \geq 1} \sum_{s \geq l} a_s p^{s-l}$$

and accordingly for b and n. Hence

$$p^t \| \binom{n}{a} = \frac{n!}{a! \cdot b!} \text{ with}$$

$$t = \sum_{l \geq 1} \sum_{s \geq l}(n_s - a_s - b_s)p^{s-l} = \sum_{l \geq 1} \sum_{s \geq l}(c_s - pc_{s+1})p^{s-l}$$

$$= \sum_{l \geq 1}\left[\sum_{s \geq l} c_s p^{s-l} - \sum_{s+1 \geq l+1} c_{s+1} p^{s+1-l}\right] = \sum_{l \geq 1} c_l$$

What a proof! □

Problem 15. *Prime factor*

$$\frac{100!}{(10!)^{10}}$$

by means of Legendre's proposition 13

Solution for checking. The occuring primes and their respective powers are
((2 17) (3 8) (5 4) (7 6) (11 9) (13 7) (17 5) (19 5) (23 4) (29 3)
(31 3) (37 2) (41 2) (43 2)
(47 2) (53 1) (59 1) (61 1) (67 1) (71 1) (73 1) (79 1) (83 1) (89 1)
(97 1)) □

Problem 16. *Generalise Lucas' proposition 14 to multinomial coefficients.*

Proposition 15 (Generalized Lucas' Theorem). *Let $n \geq 1$ and $r \geq 2$ be any natural numbers and p be any prime. We write the addition $n = a_1 + a_2 + \cdots + a_r$ with $a^{(q)} \geq 0$ in p-adic representations. The highest prime power dividing the multinomial coefficient is*

$$(2.2.3) \qquad p^t \parallel \frac{n!}{\prod_{1 \leq q \leq r} a^{(q)}!}$$

where t is the <u>sum</u> of carry-overs needed for the addition $n = \sum_{1 \leq q \leq r} a^{(q)}$ done in p-adic representation.

Proof. Let the p-adic representations for $1 \leq q \leq r$ be

$$a^{(q)} = \sum_{s \geq 0} a_s^{(q)} p^s \text{ with } 0 \leq a_s^{(q)} < p \text{ for all } s \geq 0$$

$$n_q = \sum_{s \geq 0} n_s p^s \text{ with } 0 \leq n_s < p \text{ for all } s \geq 0$$

Let $c_0 = 0$ and $c_s \geq 0$ be the carry-overs for the p-adic addition $n = \sum_{1 \leq q \leq r} a^{(q)}$. Note that for $q \geq 3$ carry-overs larger than one may occur. According to the algorithm for the p-adic addition

2.2. MORE ABOUT PRIMES

holds for all $s \geq 0$

$$m_s = \sum_{1 \leq q \leq r} a_s^{(q)} + c_s$$

$$n_s = \begin{cases} m_s & \text{if } m_s < p; \\ m_s - p\lfloor \frac{m_s}{p} \rfloor & \text{if } m_s \geq p \end{cases} \quad \text{and}$$

$$c_{s+1} = \begin{cases} 0 & \text{if } m_s < p; \\ \lfloor \frac{m_s}{p} \rfloor & \text{if } m_s \geq p \end{cases}$$

$$n_s = \sum_{1 \leq q \leq r} a_s^{(q)} + c_s - pc_{s+1}$$

The prime powers in the factorials are according to formula (2.2.1) from Legendre's proposition 13

$$p^{\alpha^{(q)}} \| a^{(q)}! \quad \text{with} \quad \alpha^{(q)} = \sum_{l \geq 1} \lfloor \frac{a^{(q)}}{p^l} \rfloor = \sum_{l \geq 1} \sum_{s \geq l} a_s^{(q)} p^{s-l}$$

and accordingly for n. Hence holds

$$p^t \| \frac{n!}{\prod_{1 \leq q \leq r} a^{(q)}!} \quad \text{with}$$

$$t = \sum_{l \geq 1} \sum_{s \geq l} \left(n_s - \sum_{1 \leq q \leq r} a_s^{(q)} \right) p^{s-l} = \sum_{l \geq 1} \sum_{s \geq l} (c_s - pc_{s+1}) p^{s-l}$$

$$= \sum_{l \geq 1} \left[\sum_{s \geq l} c_s p^{s-l} - \sum_{s+1 \geq l+1} c_{s+1} p^{s+1-l} \right] = \sum_{l \geq 1} c_l$$

What a proof! □

2.2.2 Stirling's Formula

Lemma 3. *The sequence*

(2.2.4) $$a_n := \frac{n!}{n^n e^{-n} \sqrt{n}}$$

with $n \geq 1$ is decreasing.

Proof. Let $n \geq 1$. Equivalent are

$$a_n > a_{n+1}$$
$$\frac{n!}{n^{(n+\frac{1}{2})}e^{-n}} > \frac{(n+1)!}{(n+1)^{(n+\frac{3}{2})}e^{-n-1}}$$
$$(n+1)^{(n+\frac{3}{2})}e^{-n-1} > (n+1)n^{(n+\frac{1}{2})}e^{-n}$$
$$(n+1)^{(n+\frac{1}{2})} > n^{(n+\frac{1}{2})}e$$
$$\left(1+\tfrac{1}{n}\right)^{(n+\frac{1}{2})} > e$$
$$\log\left(1+\tfrac{1}{n}\right) > \tfrac{2}{2n+1}$$

To check the last inequality, we use the series expansion of the logarithm

$$\log(1+x) = x - \frac{x^2}{2} + \frac{x^3}{3} - \frac{x^4}{4} \pm \ldots$$

which gives for $|x| < 1$ alternating upper and lower bounds. Hence it is sufficient to check

$$\left(\frac{1}{n} - \frac{1}{2n^2} + \frac{1}{3n^3} - \frac{1}{4n^4}\right) \geq \frac{2}{2n+1}$$

which is in turn equivalent to

$$(2n+1)(12n^3 - 6n^2 + 4n - 3) \geq 24n^4$$
$$24n^4 - 12n^3 + 8n^2 - 6n + 12n^3 - 6n^2 + 4n - 3 \geq 24n^4$$
$$2n^2 - 2n - 3 \geq 0$$

which hold for $n \geq 2$. Checking backwards completes the proof.

What happens for $n = 1$? Clearly, the equivalence

$$a_n > a_{n+1} \Leftrightarrow \log\left(1+\frac{1}{n}\right) > \frac{2}{2n+1}$$

still remains true for $n = 1$. Too, one may directly check the inequality $e = a_1 > a_2$ to hold. But the series expansion of $\log(1 + x)$ is only convergent for $-1 < x \leq 1$ and thus converges very slowly for $x = 1$.. □

Problem 17. *How many terms of the logarithmic series are needed to confirm* $\log 2 > \frac{2}{3}$*?*

2.2. MORE ABOUT PRIMES

Solution. One really needs at least 20 terms:

$$\log 2 > \sum_{1 \leq k \leq 20} \frac{(-1)^{k-1}}{k} > \frac{2}{3}$$

□

Theorem 3 (Wallis' product).

(2.2.5) $$\frac{\pi}{2} = \lim_{m \to \infty} \frac{16^m (m!)^4}{(2m)!(2m+1)!}$$

Proof. Let $n \geq 2$. Integration by parts yields

$$\int_0^\pi \sin^n x \, dx = \left[-\cos x \sin^{n-1} x\right]_0^\pi + (n-1) \int_0^\pi \cos^2 x \sin^{n-2} x \, dx$$

$$= (n-1) \int_0^\pi (1 - \sin^2 x) \sin^{n-2} x \, dx$$

$$= (n-1) \int_0^\pi \sin^{n-2} x \, dx - (n-1) \int_0^\pi \sin^n x \, dx$$

$$\int_0^\pi \sin^n x \, dx = \frac{n-1}{n} \int_0^\pi \sin^{n-2} x \, dx$$

for all $n \geq 2$. Iterating for even $n = 2m$ yields

$$\int_0^\pi \sin^{2m} x \, dx = \frac{2m-1}{2m} \cdot \frac{2m-3}{2m-2} \cdots \frac{1}{2} \int_0^\pi \sin^0 x \, dx = \frac{(2m)!}{4^m (m!)^2} \pi$$

Iterating for odd $n = 2m+1$ yields

$$\int_0^\pi \sin^{2m+1} x \, dx = \frac{2m}{2m+1} \cdot \frac{2m-2}{2m-1} \cdots \frac{2}{3} \int_0^\pi \sin x \, dx$$

$$= \frac{4^m (m!)^2}{(2m+1)!} \cdot 2$$

The quotient of the two formula is

(2.2.6) $$\frac{\int_0^\pi \sin^{2m+1} x \, dx}{\int_0^\pi \sin^{2m} x \, dx} = \frac{16^m (m!)^4}{(2m)!(2m+1)!} \cdot \frac{2}{\pi}$$

Because of the estimate

$$\frac{2m}{2m+1} = \frac{\int_0^\pi \sin^{2m+1} x \, dx}{\int_0^\pi \sin^{2m-1} x \, dx} \leq \frac{\int_0^\pi \sin^{2m+1} x \, dx}{\int_0^\pi \sin^{2m} x \, dx} \leq 1$$

the quotient (2.2.6) has limit 1 for $m \to \infty$. Hence

$$\frac{\pi}{2} = \lim_{m \to \infty} \frac{16^m (m!)^4}{(2m)!(2m+1)!}$$

□

Corollary 6.

(2.2.7) $$\sqrt{\frac{2}{(2m+1)\pi}} \cdot 4^m < \binom{2m}{m}$$

and the quotient of the left and right-hand sides has limit 1 for $m \to \infty$.

Proof. Equation (2.2.6) implies

$$\frac{16^m (m!)^4}{(2m)!(2m+1)!} \cdot \frac{2}{\pi} < 1$$

$$\frac{2}{(2m+1)\pi} \cdot 16^m < \frac{(2m)!^2}{(m!)^4} = \binom{2m}{m}^2$$

$$\sqrt{\frac{2}{(2m+1)\pi}} \cdot 4^m < \binom{2m}{m}$$

□

Theorem 4 (Stirling's formula). *The sequence a_n is decreasing and*

(2.2.8) $$\lim_{n \to \infty} \frac{n!}{n^n e^{-n} \sqrt{n}} = \sqrt{2\pi}$$

(2.2.9) $$e \geq \frac{n!}{n^n e^{-n} \sqrt{n}} > \sqrt{2\pi} \quad \text{for all } n \geq 1$$

Proof. Since the sequence a_n is decreasing and bounded below, its limit L exists. From Wallis' product (2.2.5) we obtain

$$\frac{\pi}{2} = \lim_{m \to \infty} \frac{16^m (m!)^4}{(2m)!(2m+1)!} = \lim_{m \to \infty} \frac{16^m \cdot m^{4m+2} e^{-4m} L^4}{(2m+1)(2m)^{4m+1} e^{-4m} L^2}$$

$$= \lim_{m \to \infty} \frac{(2m)^{4m+2} L^2}{4(2m+1)(2m)^{4m+1}} = \frac{L^2}{4}$$

Hence $L = \sqrt{2\pi}$.

□

2.2.3 How Many Primes?

The number of primes $p \leq x$ is denoted by $\pi(x)$. We use the notation
$$N\sharp = \prod \{p : \text{ primes } 2 \leq p \leq N\}$$
and p_- for the prime $p < N$ preceding N. We get an upper bound for the product of primes from the observation that the middle binomial coefficient $\binom{n}{\lfloor \frac{n}{2} \rfloor}$ contains <u>all</u> primes from the interval $\lfloor \frac{n}{2} \rfloor < p \leq n$ <u>exactly once</u>.

Proposition 16 (Upper bound for the product of primes). *For any $n \geq 2$, the product of all primes less or equal n is at most 4^{n-1}:*
$$\prod \{p : \text{ primes } p \leq n\} < 4^{n-1}$$

Proof. The assertion is true for $n = 1, 2, 3$. For the induction step, we assume that the assertion is true for all products up to $n' < n$, and check the assertion for n. In the product
$$\prod \{p : \text{ primes } p \leq n\}$$
$$= \prod \{p : \text{ primes } p \leq \lfloor \frac{n}{2} \rfloor\} \cdot \prod \{p : \text{ primes } \lfloor \frac{n}{2} \rfloor < p \leq n\}$$
we use the induction assumption to estimate the first factor, and the property
$$p \parallel \binom{n}{\lfloor \frac{n}{2} \rfloor} \quad \text{for all primes } \lfloor \frac{n}{2} \rfloor < p \leq n$$
of the middle binomial coefficient to estimate the second factor. Hence we get an upper bound for the product of primes
$$\prod \{p : \text{ primes } p \leq n\} < 4^{\lfloor \frac{n}{2} \rfloor - 1} \cdot \binom{n}{\lfloor \frac{n}{2} \rfloor}$$
$$< 4^{\lfloor \frac{n}{2} \rfloor - 1} \cdot 2^n \leq 4^{\lfloor \frac{n}{2} \rfloor - 1} \cdot 4^{\lceil \frac{n}{2} \rceil} = 4^{n-1}$$
as to be shown in the induction step. □

Corollary 7.

(2.2.10) $\quad \displaystyle\sup_{\frac{N+1}{2} < p \leq N} \frac{p - p_-}{\log p} \geq \frac{N-1}{N \log 4} \quad$ *for any odd prime N.*

(2.2.11) $\quad \displaystyle\limsup_{p \to \infty} \frac{p - p_-}{\log p} \geq \frac{1}{\log 4} > .72$

Proof. The assertion is true for $N = 3$. Let $N \geq 5$ be any odd prime. Let q_- be the largest prime $q_- \leq \frac{N+1}{2}$. Hence $N - q_- \geq \frac{N-1}{2} \geq q_- - 1$

$$\frac{N-1}{2} \leq N - q_- = \sum_{\frac{N+1}{2} < p \leq N} (p - p_-)$$

$$\leq \left[\sup_{\frac{N+1}{2} < p \leq N} \frac{p - p_-}{\log p} \right] \cdot \log \left[\prod_{\frac{N+1}{2} < p \leq N} p \right]$$

$$\leq \left[\sup_{\frac{N+1}{2} < p \leq N} \frac{p - p_-}{\log p} \right] \cdot \log \left(\frac{N}{\frac{N+1}{2}} \right)$$

$$\leq \left[\sup_{\frac{N+1}{2} < p \leq N} \frac{p - p_-}{\log p} \right] \cdot N \log 2$$

$$\frac{N-1}{N \log 4} \leq \sup_{\frac{N+1}{2} < p \leq N} \frac{p - p_-}{\log p}$$

$\square$

Remark. This is only a weak preliminary result. Indeed, the prime number theorem implies that the lim sup from equation (2.2.11) is larger or equal to 1. But indeed, it is ∞. In other words, there exist unexpectedly long gaps between primes.

Proposition 17 (Upper bound for the number of primes). For all $n \geq 2$ holds

(2.2.12) $$\pi(n) \leq 2 \frac{n}{\log n}$$

Proof. The assertion is true for $n \leq 40$ as has to be checked directly. For the induction step, we assume that the assertion is true up to any $n' < 2n+1$, and check the assertion for $2n+1$.

$$\pi(2n+1) \leq \pi(n) + \frac{\log \prod \{p : \text{primes } n < p \leq 2n+1\}}{\log(n+1)}$$

$$\leq \frac{cn}{\log n} + \frac{\log \binom{2n+1}{n}}{\log(n+1)}$$

$$\leq \frac{cn}{\log n} + \frac{(2n+1)\log 2}{\log(n+1)} \leq \frac{c(2n+1)}{\log(2n+1)}$$

2.2. MORE ABOUT PRIMES

The last line holds with $c = 2$ for all $n \geq 39$. To check this assertion, define c_n by setting equality in the last line and check the following

$$\frac{(2n+1)\log 2}{\log(n+1)} = c_n \cdot \left[\frac{(2n+1)}{\log(2n+1)} - \frac{n}{\log n}\right]$$

$$\frac{(2n+1)\log 2}{\log(n+1)} = c_n \cdot \frac{(2n+1)\log n - n\log(2n+1)}{\log(2n+1)\log n}$$

$$c_n := \log 2 \cdot \frac{\log(2n+1)}{\log(n+1)} \cdot \frac{(2n+1)\log n}{(2n+1)\log n - n\log(2n+1)}$$

One now checks that the sequence c_n is decreasing for $n \geq 2$. Moreover, $c_{39} < 2$ and hence $c_n < 2$ for all $n \geq 39$ proving the claim. From $\lim_{n \to \infty} c_n = 2\log 2$, we get the corollary 8 below. $\square$

Corollary 8.

(2.2.13) $$\limsup_{n \to \infty} \pi(n)\frac{\log n}{n} \leq 2\log 2$$

2.2.4 Existence of Infinitely Many Primes

Theorem 5. *For all integers holds*

(2.2.14) $$\log N - 1 < \sum_{p \leq N} \frac{\log p}{p-1}$$

Corollary 9. *The sum*

$$\sum_{p \text{ is prime}} \frac{\log p}{p}$$

is divergent. There exist infinitely many primes.

Proof. Let $N \geq 2$ be any natural number and p be any prime. Let $p^r \parallel N!$ be the highest prime power dividing the factorial $N!$. Let L be the maximal multiplicity for which $p^L \leq N < p^{L+1}$. From Legendre's equation (2.2.1) one gets

$$r = \sum_{l \geq 1} \lfloor \frac{N}{p^l} \rfloor \leq N \cdot \left[\sum_{1 \leq l \text{ and } p^l \leq N} p^{-l}\right] = N \cdot \frac{1 - p^{-L}}{p-1} \leq \frac{N-1}{p-1}$$

Note that $p \mid N! \Leftrightarrow p \leq N$ holds for all primes p. We take the logarithm of the prime decomposition of the factorial

$$N! = \prod_{p \leq N} p^r$$

$$\log N! = \sum_{p \leq N} r \log p \leq (N-1) \sum_{p \leq N} \frac{\log p}{p-1}$$

and in the end combine with Stirling's formula (2.2.8) This formula has been shown by (2.2.9) to be a lower bound, too. For simplicity, we may in the end drop the constant term $\frac{1}{2} \log 2\pi = .91 > 0$

$$\frac{2N+1}{2} \log N - N + \tfrac{1}{2} \log 2\pi \leq \log N! \leq (N-1) \sum_{p \leq N} \frac{\log p}{p-1}$$

$$\log N - 1 < \frac{(2N+1)\log N}{2N-2} - \frac{N}{N-1} \leq \sum_{p \leq N} \frac{\log p}{p-1}$$

$\square$

Lemma 4. *Here is an upper bound corresponding to the lower bound (2.2.14). For any $N \geq 5$.holds*

$$\sum_{p \leq N} \frac{\log p}{p-1} < 2 \log N$$

Proof. One needs the upper estimate of the product of primes from the middle binomial coefficient

$$\prod \{p : \text{primes } 2^r < p < 2^{r+1}\} \leq \binom{2^{r+1}}{2^r} \leq 2^{r+1}$$

Let $N \geq 5$ and choose R such that $2^R < N \leq 2^{R+1}$. Hence $R \log 2 < \log N$.

$$\sum_{5 \leq p \leq N} \frac{1}{p-1} \leq \sum_{2 \leq r \leq R} \left[\sum_{2^r < p < 2^{r+1}} \frac{\log p}{p-1} \right]$$

$$\leq \sum_{2 \leq r \leq R} 2^{-r} \cdot \prod \{p : \text{primes } 2^r < p < 2^{r+1}\}$$

$$\leq \sum_{2 \leq r \leq R} \frac{2^{r+1} \log 2}{2^r} = 2(R-1) \log 2 \leq 2 \log N - \log 4$$

$$\sum_{p \leq N} \frac{\log p}{p-1} < 2 \log N - \log 4 + \log 2 + \frac{\log 3}{2} < 2 \log N$$

2.2. MORE ABOUT PRIMES

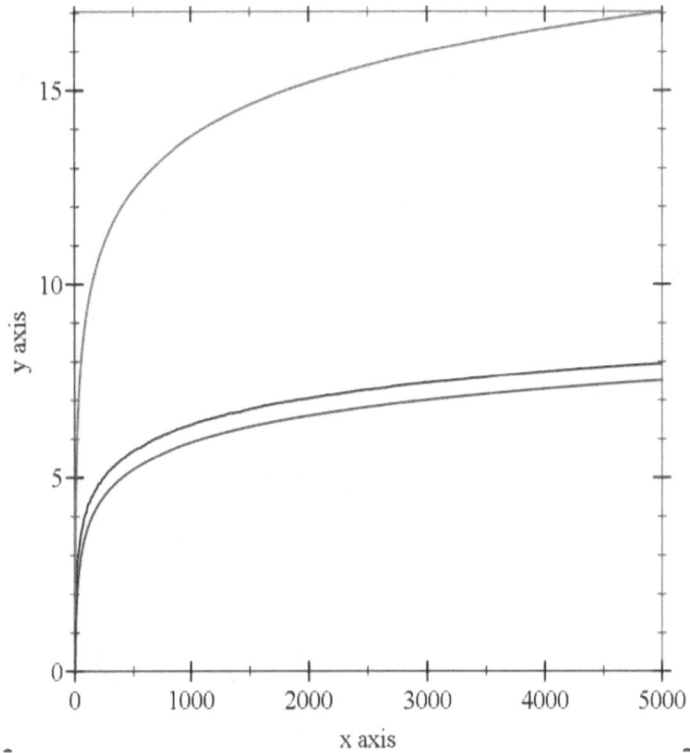

Figure 2.2.1: Upper and lower estimates for $\sum_{p\leq N} \frac{\log p}{p-1}$.

$\square$

Remark. I am not aware whether the estimates for $\sum_{p\leq N} \frac{\log p}{p-1}$ have been published anywhere.

Remark. The upper and lower estimates for $\sum_{p\leq N} \frac{\log p}{p-1}$ are calculated in the figure on page 39 for $N \leq 5\,000$. Moreover we have obtained
$$(2.2.15) \quad 1 \leq \liminf_{N\to\infty} \frac{1}{\log N} \sum_{p\leq N} \frac{\log p}{p-1} \leq \limsup_{N\to\infty} \frac{1}{\log N} \sum_{p\leq N} \frac{\log p}{p-1} \leq 2$$

I conjecture that in reality the limit exists and is equal to 1.

Conjecture 1. *The limit*

(2.2.16) $$\lim_{N\to\infty} \sum_{p\leq N} \frac{\log p}{p-1} - \log N \approx -.58$$

exists. I have checked $N \leq 40\,000$.

Theorem 6 (Euler's infinitely many primes).

$$\sum_{p\leq N} \frac{1}{p} > \log(\log N) - \frac{1}{4}$$

for any $N \geq 2$.

Corollary 10. *The sum*

$$\sum_{p \text{ is prime}} \frac{1}{p}$$

is divergent. There exist infinitely many primes.

For the proof we shall use the following facts:

(i)
$$\log(1-x)^{-1} \leq x + \frac{x^2}{2} \quad \text{for } -1 \leq x < 1$$

as one gets from the series expansion of the logarithm.

(ii)
$$\sum_{1\leq n\leq N} \frac{1}{n} > \int_1^{N+1} \frac{dn}{n} = \log(N+1)$$

(iii)
$$\sum \frac{1}{p^2} < \frac{1}{2}$$

Here is a reason:

$$\sum \frac{1}{p^2} < \frac{1}{4} - 1 + \sum_{k\geq 1} \frac{1}{(2k-1)^2} = -\frac{3}{4} + \frac{\pi^2}{8} < -\frac{3}{4} + \frac{10}{8} = \frac{1}{2}$$

Proof. The main secret is to distribute the following product over primes

(2.2.17)
$$\prod_{p\leq N} \left(1 - \frac{1}{p}\right)^{-1} = \sum \{\frac{1}{n} : n \text{ has only prime factors } p \leq N\}$$

2.2. MORE ABOUT PRIMES

We neglect the terms with $n > N$ and take the logarithm of the resulting estimate

$$\sum_{n \leq N} \frac{1}{n} \leq \prod_{p \leq N} \left(1 - \frac{1}{p}\right)^{-1}$$

$$\log \sum_{n \leq N} \frac{1}{n} \leq \log \prod_{p \leq N} \left(1 - \frac{1}{p}\right)^{-1}$$

$$\log \log (N+1) \leq \sum_{p \leq N} \log \left(1 - \frac{1}{p}\right)^{-1} \quad \text{from remark (ii)};$$

$$\log \log (N+1) \leq \sum_{p \leq N} \frac{1}{p} + \frac{1}{2} \sum_p \frac{1}{p^2} \leq \frac{1}{4} + \sum_{p \leq N} \frac{1}{p}$$

by remarks (i) and (iii). $\square$

Problem 18. *Prove the main formula (2.2.17) by induction on the number of primes.*

I give an upper bound corresponding to Euler's result.

Lemma 5. $\sum_{p \leq N} \frac{1}{p} < 1.6 + 2 \log \log N$ *for any* $N \geq 2$.

Proof. One needs the upper estimate of the number of primes from the middle binomial coefficient

$$\pi(2^{r+1}) - \pi(2^r) \leq [r \log 2]^{-1} \log \left[\prod \{p : \text{primes } 2^r < p < 2^{r+1}\}\right]$$

$$\leq \frac{\log \binom{2^{r+1}}{2^r}}{r \log 2} \leq \frac{2^{r+1} \log 2}{r \log 2} = \frac{2^{r+1}}{r}$$

Let $N \geq 5$ and choose R such that $2^R < N \leq 2^{R+1}$. Hence

$R \log 2 < \log N$.

$$\sum_{p \le N} \frac{1}{p} \le \frac{5}{6} + \sum_{2 \le r \le R} \left[\sum_{2^r < p < 2^{r+1}} \frac{1}{p} \right]$$

$$\le \frac{5}{6} + \sum_{2 \le r \le R} \frac{\pi(2^{r+1}) - \pi(2^r)}{2^r + 1} \le \frac{5}{6} + \sum_{2 \le r \le R} \frac{2}{r}$$

$$\le \frac{5}{6} + \int_1^R \frac{2 dr}{r} = \frac{5}{6} + 2 \log R \le \frac{5}{6} + 2 \log \frac{\log N}{\log 2}$$

$$= \frac{5}{6} + 2 \log \log N - 2 \log (\log 2)$$

$$\sum_{p \le N} \frac{1}{p} < 1.6 + 2 \log \log N$$

$\square$

Remark. The upper and lower estimates for $\sum_{p \le N} \frac{1}{p}$ are calculated in the figure on page 43. Moreover we have obtained
(2.2.18)
$$1 \le \liminf_{N \to \infty} (\log \log N)^{-1} \sum_{p \le N} \frac{1}{p} \le \limsup_{N \to \infty} (\log \log N)^{-1} \sum_{p \le N} \frac{1}{p} \le 2$$

I conjecture that in reality the limit exists and is equal to 1.

Conjecture 2. *The limit*

(2.2.19) $$\lim_{N \to \infty} \sum_{p \le N} \frac{1}{p} - \log \log N \approx .26$$

exists. I have checked $N \le 10\,000$.

Proposition 18 (Partial summation). *Let $a_1, \ldots, a_n$ and $b_1, \ldots, b_n$ be any finite sequences from a ring $\mathcal{R}$. Put $A_0 := 0$ and*

$$A_k = \sum_{1 \le i \le k} a_i \quad \text{for all } 1 \le k \le n$$

The partial summation formula is

(2.2.20) $$\sum_{1 \le k \le n} a_k b_k = A_n b_n - \sum_{2 \le k \le n} A_{k-1}(b_k - b_{k-1})$$

2.2. MORE ABOUT PRIMES

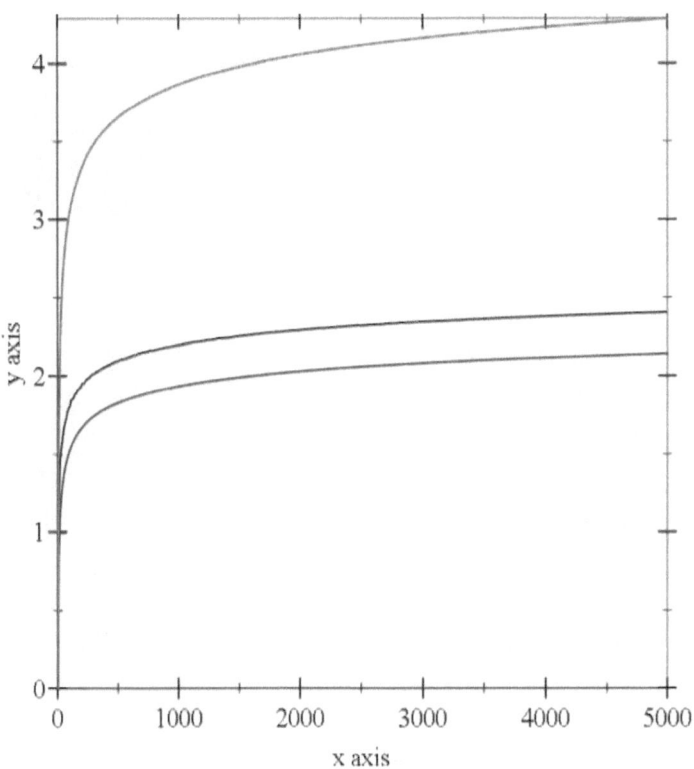

Figure 2.2.2: Upper and lower estimates for $\sum_{p \leq N} \frac{1}{p}$.

Proof. We use induction by n. For $n = 1$ we just state $a_1 b_1 = A_1 b_1$. For $n = 2$ we just state $a_1 b_1 + a_2 b_2 = A_2 b_2 - A_1(b_2 - b_1)$. Here is the induction step $n \to n+1$:

$$\sum_{1 \le k \le n+1} a_k b_k = \sum_{1 \le k \le n} a_k b_k + a_{n+1} b_{n+1}$$

$$= A_n b_n - \sum_{2 \le k \le n} A_{k-1}(b_k - b_{k-1}) + a_{n+1} b_{n+1}$$

$$= A_n b_{n+1} - A_n(b_{n+1} - b_n)$$
$$- \sum_{2 \le k \le n} A_{k-1}(b_k - b_{k-1}) + a_{n+1} b_{n+1}$$

$$= A_n b_{n+1} - \sum_{2 \le k \le n+1} A_{k-1}(b_k - b_{k-1}) + a_{n+1} b_{n+1}$$

$$= A_{n+1} b_{n+1} - \sum_{2 \le k \le n+1} A_{k-1}(b_k - b_{k-1})$$

$\square$

Problem 19. *For simplicity assume that N is a prime. Prove the formula*

$$\sum_{p \le N} \frac{\log p}{p} = \frac{\log N\sharp}{N} + \sum_{3 \le p \le N} \frac{(\log p_-\sharp) \cdot (p - p_-)}{p \cdot p_-}$$

Lemma 6.

$$\frac{\log N}{\log 4} \le 1.14 + \sum_{3 \le p \le N} \frac{p - p_-}{p - 1}$$

for any prime N.

Proof. We use partial summation for the second of formulas (2.2.14) and use on the right-hand side the upper bound for the

2.2. MORE ABOUT PRIMES

product of primes from proposition 16:

$$\frac{(2N-1)\log N}{2N} \leq \sum_{p \leq N} \frac{\log p}{p-1}$$

$$= \frac{\log N\sharp}{N-1} + \sum_{3 \leq p \leq N} \frac{(\log p_-\sharp) \cdot (p - p_-)}{(p-1)(p_- - 1)}$$

$$\leq \frac{(N-1)\log 4}{N-1} + \log 4 \sum_{3 \leq p \leq N} \frac{(p_- - 1) \cdot (p - p_-)}{(p-1)(p_- - 1)}$$

$$\log N \leq \frac{\log N}{2N} + \log 4 + \log 4 \sum_{3 \leq p \leq N} \frac{p - p_-}{p - 1}$$

Finally $N < 2^N$ implies $\log N < N \log 2$. Hence one gets the result as claimed. $\square$

2.2.5 Bertrand's Postulate

Theorem 7 (Bertrand's postulate). *Between any number $n \geq 2$ and its double exists a prime.*

Remark. It is easy to verify Bertrand's postulate up to any small number N. To this end it is enough to produce a finite sequence of primes $2, 3, \ldots p_k, p_{k+1}$ such that $p_k \geq N$ and $p_i < 2p_{i-1}$ for all indexes i. For example because of the sequence of primes

$$2, 3, 5, 7, 13, 23, 43, 83, 163, 317$$

we see [1] that Bertrand's postulate holds for $n \leq 163$.

This paragraph takes some inspiration from the book [1] by Martin Aigner and Günter M. Ziegler.

Lemma 7. *Let $n \geq 2$. The middle binomial coefficient has the prime decomposition*

(2.2.21)
$$\binom{2n}{n} = P_1 \cdot P_2 \cdot P_3$$

$$= \prod \{p^{r-1} : p^r \| \binom{2n}{n} \text{ and } r \geq 2\}$$

$$\cdot \prod \{p : p \mid \binom{2n}{n} \text{ and } 2 \leq p \leq \tfrac{3n}{2}\} \cdot \prod \{p : n < p < 2n\}$$

where the symbol p stands for any prime.

---

[1] after having checked that $323 = 17 * 19, 319 = 11 * 29$ are composite

Factor P_1 contains only primes $p \leq \sqrt{2n-1}$. Moreover

$$r \leq d = \left\lfloor \frac{\log_2 n}{\log_2 p} \right\rfloor + 1$$

Factor P_2 contains any primes $2 \leq p \leq \frac{2n}{3}$ at most once. For any prime $\frac{2n}{3} < p \leq n$ holds $p \nmid \binom{2n}{n}$.

Factor P_3 contains <u>all</u> primes $n < p < 2n$ <u>exactly once</u>. For any prime $n < p \leq 2n$ holds $p \| \binom{2n}{n}$.

Proof. **About factor P_1:** Let $p^r \| \binom{2n}{n}$. By Lucas' proposition 14, the multiplicity r of any prime p equals the number of carry-overs needed for the addition $2n = n + n$ done in p-adic representation. Surely $r \leq d$ which is the number of digits needed for the p-adic representation of $2n$. Multiple primes with $r \geq 2$ are only possible if $d \geq 3$ since only a <u>third</u> digit for $2n$ may cause two carry-overs. Now $d \geq 3$ is equivalent to $\log 2n \geq 2 \log p$, which is in turn equivalent to $p \leq \sqrt{2n}$ and indeed $p \leq \sqrt{2n-1}$.

About factor P_2: For any prime $\frac{2n}{3} < p \leq n$ holds $p \nmid \binom{2n}{n}$. Here is the reason:

$p^2 \| (n!)^2$ since p (but not $2p$) is the only factor occurring in $n!$ which divisible by p.

$p^2 \| (2n)!$ since p and $2p$ (but not $3p$) are the only two factors occurring in $(2n)!$ which are divisible by p.

About factor P_3: For any prime $n < p \leq 2n$ holds $p \| \binom{2n}{n}$. Here is the reason:

$p \nmid n!$ but $p \| (2n)!$ since p (but not $2p$) is the only factor in $(2n)!$ divisible by p.

$\square$

We use lemma 7 to get a lower estimate of the third factor P_3. To this end, we need upper bounds for the factors P_1 and P_2..

Lemma 8. *For $n \geq 33$ holds*

$$\log P_1 = \log \left[\prod \{ p^{r-1} : p^r \| \binom{2n}{n} \text{ and } r \geq 2 \} \right] \leq \min[8, \log n] \frac{\sqrt{2n}}{2}$$

2.2. MORE ABOUT PRIMES

Proof. To estimate an upper bound or the number of primes, we use proposition 17 above.

$$\log P_1 = \sum \{(r-1)\log p \ : \ p^r \ \| \ \binom{2n}{n} \text{ and } r \geq 2\}$$
$$\leq \sum \{\left\lfloor \frac{\log n}{\log p} \right\rfloor \log p \ : \ p^2 \ | \ \binom{2n}{n} \text{ and } p \leq \sqrt{2n}\}$$
$$\leq \sum \{\log n \ : \ p \leq \sqrt{2n}\} \leq \pi(\sqrt{2n})\log n$$
$$\leq \min[8, \log n]\frac{\sqrt{2n}}{2}$$

$\square$

Remark. The simple estimate for the number of primes up to $x \geq 8$ is $\pi(x) \leq \frac{x}{2}$. We may assume $n \geq 33$ and hence $\sqrt{2n-1} \geq 8$ and get from this estimate

$$\pi\left(\sqrt{2n-1}\right)\log n \leq \log n \frac{\sqrt{2n}}{2}$$

Alternatively, we may use the upper bound for the number of primes from proposition (17) and get

$$\pi\left(\sqrt{2n-1}\right)\log n < 2\frac{\sqrt{2n}}{\log\sqrt{2n}}\log n < 8\frac{\sqrt{2n}}{2}$$

The first simpler way is better for $\log n \leq 8$ hence $n \leq 2980$.

Lemma 9.

$$\log P_2 = \log\left[\prod\{p \ : \ p \ | \ \binom{2n}{n} \text{ and } 2 \leq p \leq \frac{2n}{3}\}\right] \leq \frac{2n-3}{3}\log 4$$

Lemma 10. Let $n \geq 2$.

(2.2.22)
$$\log\left[\prod\{p \ : \ n < p < 2n\}\right] > \frac{n}{3}\log 4 - \min[8, \log n]\frac{\sqrt{2n}}{2} - \frac{1}{2}\log n$$

Proof.

$$\log\left[\prod\{p : n < p < 2n\}\right] \geq \log\binom{2n}{n} - \log P_1 - \log P_2$$

$$\geq \log\left[\sqrt{\tfrac{2}{(2n+1)\pi}} \cdot 4^n\right] - \min[8, \log n]\tfrac{\sqrt{2n}}{2} - \tfrac{2n-3}{3}\log 4$$

$$\geq \tfrac{n}{3}\log 4 - \min[8, \log n]\tfrac{\sqrt{2n}}{2} - \log\sqrt{\tfrac{(2n+1)\pi}{2}} + \log 4$$

$$= \tfrac{n}{3}\log 4 - \min[8, \log n]\tfrac{\sqrt{2n}}{2} - \log\sqrt{n} - \log\sqrt{\tfrac{2n+1}{2n}}$$

$$\quad - \log\sqrt{2\pi} + \tfrac{1}{2}\log 2 + 2\log 2$$

$$> \tfrac{n}{3}\log 4 - \min[8, \log n]\tfrac{\sqrt{2n}}{2} - \tfrac{1}{2}\log n$$

since the four last terms together are positive:

$$-\log\sqrt{\tfrac{2n+1}{2n}} - \log\sqrt{2\pi} + \tfrac{5}{2}\log 2 > -\tfrac{1}{4n} - 1 + \tfrac{5}{3} = \tfrac{2}{3} - \tfrac{1}{4n} > 0$$

□

Corollary 11. *For all $n \geq 2$ there exists a prime $n < p < 2n$.*

Proof. The assertion holds for $n \leq 163$ by the remark 2.2.5 made at the beginning. Now assume $n \geq 33$ and use estimate (2.2.22) with $n = 2^r$. We get

$$6\log\left[\prod\{p : n < p < 2n\}\right] > 2n\log 4 - 3\sqrt{2n}\log n - 3\log n$$

$$= (4n - 3r\sqrt{2n} - 3r)\log 2 =: K\log 2$$

Put $r := 5$, $n = 32$ and see it is too small since
$$K = 4n - 3r\sqrt{2n} - 3r > 128 - 15*8 - 15 < 0$$
Put $r := 6$, $n = 64$ and get
$$K = 4n - 3r\sqrt{2n} - 3r > 256 - 18*8*3/2 - 18 = 22 > 0$$

Hence $\prod\{p : n < p < 2n\} > 1$ for $n \geq 64$ which proves existence of a prime n the interval $n < p < 2n$. □

Corollary 12. *In the limit $n \to \infty$, the number of primes in the interval $n < p < 2n$ are bounded below by* [2]

$$\liminf_{n\to\infty}[\pi(2n) - \pi(n)]\frac{\log n}{n} \geq \frac{4\log 2}{3} > \frac{8}{9}$$

---

[2]The true limit is 1.

2.2. MORE ABOUT PRIMES

Proof. $L = \pi(2n) - \pi(n)$ primes exist in the interval $n < p < 2n$.

$$[\pi(2n) - \pi(n)] \log(2n) > \log\left[\prod\{p : n < p < 2n\}\right]$$
$$> \frac{n}{3}\log 4 - 4\sqrt{2n} - \frac{\log n}{2}$$
$$[\pi(2n) - \pi(n)]\frac{\log(2n)}{n} > \frac{4\log 2}{3} - \frac{8}{\sqrt{2n}} - \frac{\log n}{2n}$$

Since the last two terms have limit zero holds for $n \to \infty$

$$\liminf_{n\to\infty}[\pi(2n) - \pi(n)]\frac{\log n}{n} \geq \frac{4\log 2}{3} > \frac{8}{9}$$

□

Corollary 13.

(2.2.23) $$\liminf_{n\to\infty} \pi(n)\frac{\log n}{n} \geq \frac{4\log 2}{3} > \frac{8}{9}$$

Proof. Let

$$b_n := \pi(n)\frac{\log n}{n} \text{ and } b := \liminf_{n\to\infty} b_n$$

Take limit $n \to \infty$ in a subsequence where $b_{2n} \to b$ and $b_n \to b$ both converge.

$$\pi(2n)\frac{\log 2n}{2n} = \left[[\pi(2n) - \pi(n)]\frac{\log n}{2n} + \pi(n)\frac{\log n}{2n}\right]\frac{\log 2n}{\log n}$$

$$\liminf_{n\to\infty} \pi(2n)\frac{\log 2n}{2n} \geq \liminf_{n\to\infty}[\pi(2n) - \pi(n)]\frac{\log n}{2n}$$
$$+ \liminf_{n\to\infty} \pi(n)\frac{\log n}{2n}$$
$$\geq \frac{2\log 2}{3} + \frac{b}{2} \text{ hence } b \geq \frac{4\log 2}{3}$$

□

Proposition 19 (Lower bound for the product of primes).
For any $n \geq 2$, the logarithmic product of all primes $p \leq n$ is at least

(2.2.24) $$\log\prod\{p : \text{primes } p \leq n\}$$
$$\geq 9.5 + \frac{n\log 4}{3} - (4+\sqrt{8})\sqrt{n} - \frac{(\log n)^2}{\log 16}$$

Proof. As an induction start, the assertion has to be checked directly for $n = 2$. Because of a difficulty appearing below, one has directly to check that the assertion (2.2.24) holds for $2 \leq n \leq 67$. This claim is rather obvious since the left-hand side is increasing, but the right-hand side is decreasing for $n \leq 67$.

We now assume $n \geq 34$. For the induction step, we assume that the assertion is true for all products up to $n' < 2n - 1$ (which is $2n - 1 \geq 2 * 34 - 1 = 67$), and check the assertion for $2n - 1$ and $2n$. In the logarithmic product

$$\log \prod \{p : \text{primes } p \leq 2n - 1\}$$
$$= \log \prod \{p : \text{primes } p \leq n\} + \log \prod \{p : \text{primes } n < p \leq 2n - 1\}$$

we use the induction assumption to estimate the first summand and lemma 10 to estimate the second factor. Hence we get a lower bound for the logarithmic product of primes up to $2n$ from the estimate

$$\log(2n)\sharp = \log(2n - 1)\sharp$$
$$\geq D + An \log 4 - B\sqrt{2n} - C(\log n)^2$$
$$+ \tfrac{n}{3} \log 4 - 4\sqrt{2n} - \tfrac{1}{2} \log n$$
$$\geq D + 2An \log 4 - B\sqrt{4n} - C(\log (2n))^2$$

To this end, we choose the constants A, B, C as follows

$$A \leq \tfrac{1}{3}, \quad \text{we put } A = \tfrac{1}{3}$$
$$B(\sqrt{2} - 1) \geq 4, \quad \text{we put } B = 2(\sqrt{2} + 1)$$
$$C((\log (2n))^2 - (\log n)^2) \geq C2 \log n \log 2 \geq \tfrac{1}{2} \log n,$$
$$\text{we put } C = \tfrac{1}{4 \log 2}$$

as to complete the induction step up to $2n$. Next one checks that

$$A \log 4 \geq B(\sqrt{4n} - \sqrt{4n - 2}) + C[(\log (2n))^2 - (\log (2n - 1))^2]$$

holds for $n = 34$. Since the right-hand side is decreasing, the inequality remains true for all $n \geq 34$. Hence

$$\log(2n)\sharp = \log(2n - 1)\sharp \geq 2An \log 4 - B\sqrt{4n} - C(\log (2n))^2$$
$$\geq (2n - 1)A \log 4 - B\sqrt{4n - 2} - C(\log (2n - 1))^2$$

which completes the induction step up to $2n - 1$. □

2.2. MORE ABOUT PRIMES

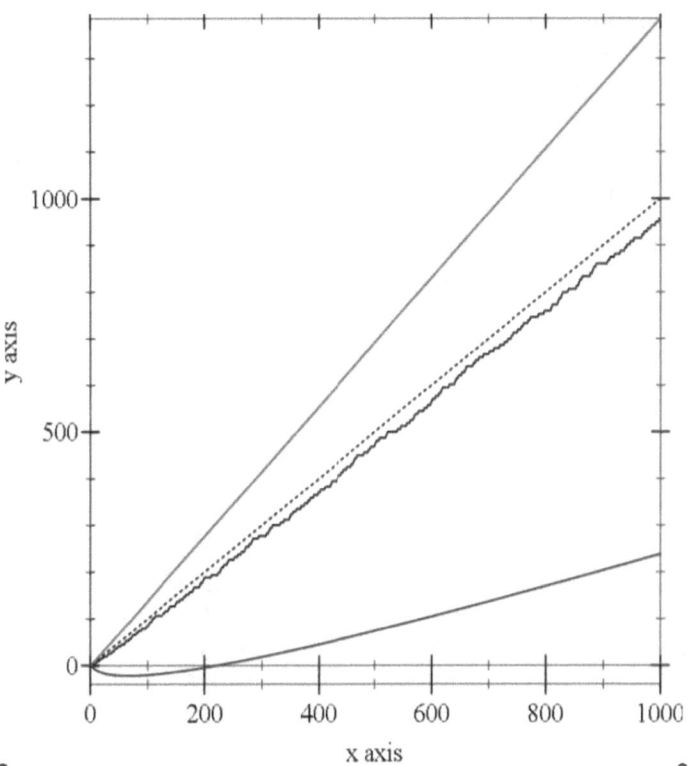

Figure 2.2.3: Our upper and lower estimates for $\log n\sharp$ are rather poor.

Remark. Since the right-hand side has a minimum at $n = 67$ but the left-hand side is increasing for $n \geq 2$, it is impractical to use the estimate (2.2.24) for n less than about 67. Indeed we see from the figure on page 51 that the lower bound is even negative for $n \leq 222$. Too, one sees from the figure how far the gap between the lower bound (blue curve) and the upper bound (red curve) for the logarithmic product of primes from proposition 16 is. The black wiggled curve is the exact value of $\log n\sharp$, begging for an exact estimate of its wiggles.. The methods of this section are just too weak to get a really satisfactory result,—beyond the proof of Bertrand's postulate.

2.2.6 Infinitely Many Primes under Restrictions

Euclid's theorem 2 telling that there exist infinitely many primes can be reproved and sharpened in many ways. Here are some easy ones.

Proposition 20. *There exist infinitely many primes* $p \equiv -1 \pmod{4}$.

Proof. Put the first k primes $p \equiv -1 \pmod 4$ into the increasing sequence p_i starting as $3, 7, 11, \ldots$ and let

$$P = \prod_{1 \leq i \leq k} p_i$$

We may factor the number $4P - 1$. It is impossible that all prime factors of $4P - 1$ are $q_j \equiv 1 \pmod 4$ since $4P - 1 \equiv -1 \pmod 4$. Hence there exists a prime $q \mid 4P - 1$ satisfying $q \equiv -1 \pmod 4$. Too, the prime q is different from all p_i with $1 \leq i \leq k$. We conclude that there exist at least $k + 1$ primes $\equiv -1 \pmod 4$. Since this argument holds for all natural numbers k, the number of primes $\equiv -1 \pmod 4$ is not finite. $\square$

Lateron, we get as a consequence of corollary 32 that there exist infinitely many primes $p \equiv +1 \pmod 4$.

Problem 20. *Prove that there exist infinitely many primes $p \equiv 2 \pmod 3$.*

Solution. Put the first k primes $p \equiv 2 \pmod 3$ into the increasing sequence p_i starting as $5, 11, 17, 23, \ldots$ and let

$$P = \prod_{1 \leq i \leq k} p_i$$

2.2. MORE ABOUT PRIMES

We may factor the number $3P+2$. Clearly, this number is odd and not divisible by 3. Hence all its prime factors are $\equiv 1 \pmod 3$ or $\equiv 2 \pmod 3$ It is impossible that all prime factors of $3P+2$ are $q_j \equiv 1 \pmod 3$ since $3P+2 \equiv 2 \pmod 3$. Hence there exists a prime $q \mid 3P+2$ satisfying $q \equiv 2 \pmod 3$. Too, the prime q is different from all p_i with $1 \leq i \leq k$. We conclude that there exist at least $k+1$ primes $\equiv 2 \pmod 3$. Since this argument holds for all natural numbers k, the number of primes $\equiv 2 \pmod 3$ is not finite. □

Problem 21. *Try to prove that there exist infinitely many primes $p \equiv 1 \pmod 3$.*

Just try . Put the first k primes $p \equiv 1 \pmod 3$ into the increasing sequence p_i starting as $7, 13, 19, 29 \ldots$ and let

$$P = \prod_{1 \leq i \leq k} p_i$$

We may factor the number $6P+1$. Clearly, this number is odd and not divisible by 3. Hence all its prime factors are either $\equiv 1 \pmod 3$ or $\equiv 2 \pmod 3$. It is <u>possible</u> that all prime factors of $6P+1$ are $q_i \equiv 1 \pmod 3$ but <u>possible as well</u> that all prime factors of $6P+1$ are $q_i \equiv 2 \pmod 3$. Now what? □

Remark. I give a different solution in proposition 68 below.

2.2.7 Infinitely Many Composite Numbers

Proposition 21. *A nonconstant integer polynomial assumes infinitely many composite values $P(n)$, for appropriately chosen natural numbers n.*

Proof. Let $d \geq 1$ be the degree of the given polynomial. By the fundamental theorem of algebra the polynomial P can assume any value at most d times. [3] Hence there exist at most $3d$ natural numbers n for which $P(n)$ is either $0, -1$ or 1. Let a be natural number such that $P(a) \neq 0, \pm 1$. Let p be any prime factor of $P(a)$ and put $P(a) = pq$.

---

[3]Indeed, we only need to know for this proof that any value can be assumed <u>at most</u> as many times as the degree of the polynomial. This fact is stated below as Lagrange's Theorem 11.

It is easy to see that for all integer k, the difference $P(a+kp) - P(a)$ is divisible by p. Indeed, this follows by the binomial theorem. For any given polynomial with integer coefficients c_l

$$P(x) = \sum_{l=0}^{d} c_l x^l \quad \text{implies}$$

$$P(a+kp) - P(a) = \sum_{l=0}^{d} c_l \left[(a+kp)^l - a^l\right]$$

$$= \sum_{l=0}^{d} c_l \sum_{j=1}^{l} \binom{l}{j} a^{l-j}(kp)^j$$

$$= p \cdot \left[\sum_{l=0}^{d} c_l \sum_{j=1}^{l} \binom{l}{j} a^{l-j} k^j p^{j-1}\right]$$

$$P(a+kp) - P(a) = p \cdot m$$

where m is an integer abbreviating the last double sum. Since $P(a) = pq$, we conclude $P(a+kp) = p(m+q)$ is divisible by p for all integer k.

There exist at most $3d$ integers k for which $P(a+kp)$ is either 0, $-p$ or p. Hence there exist infinitely many integers k such that $a + kp \geq 0$ and $P(a+kp) \neq 0, \pm p$ is divisible by p and hence composite. □

Proposition 22. *There exist arbitrary long sequences $N, N+1, \ldots, N+l$ consisting only of composite numbers.*

Proof. Let $k \geq 2$. Put the primes into the increasing sequence p_i and let

$$P = \prod_{1 \leq i \leq k} p_i$$

The numbers

$$P+2,\ P+3,\ P+4,\ P+5,\ P+6, \ldots, P+p_k,\ P+p_k+1$$

are p_k successive numbers which are all composite. Take any $2 \leq d \leq p_k$. There exists $1 \leq i \leq k$ such that $p_i \mid d$ and hence $p_i \mid P+d$ and $p_i < P+d$. One gets the respective proper divisors

$$2|P+2,\ 3|P+3,\ 2|P+4,\ 5|P+5,\ 6|P+6, \ldots$$
$$p_k|P+p_k,\ 2|P+p_k+1$$

We conclude that $P+d$ is composite. □

2.2. MORE ABOUT PRIMES

Remark. Let $k \geq 3$. Too, the numbers

$$2 \mid P-2, 3 \mid P-3, 2 \mid P-4, 5 \mid P-5, 6 \mid P-6, \ldots$$
$$p_k \mid P - p_k, 2 \mid P - p_k - 1$$

are p_k successive numbers which are all composite. Indeed take any number $2 \leq d \leq p_k$. There exists a prime $p_i \mid d$ among $1 \leq i \leq k$. We conclude $p_i \mid P - d$ and for $k \geq 3$ holds $p_i < 5p_i \leq 6p_i - d \leq P - d$. Hence $P - d$ is composite.

Remark. Under the premisses that both $P - 1$ and $P + 1$ are composite, one would get a gap of double length. That would be surprising. But the proved gap has no surprising length. As one sees from the following corollary, we have only confirmed the former corollary 7.

Corollary 14.

(2.2.25) $$\limsup_{p \to \infty} \frac{p - p_-}{\log p} \geq \frac{1}{\log 4} > .72$$

Proof.

$$\limsup_{p \to \infty} \frac{p - p_-}{\log p_-} \geq \limsup_{k \to \infty} \frac{\mathcal{N}ext(P + p_k + 1) - (P + 1)_-}{\log(P + 1)_-}$$
$$\geq \limsup_{k \to \infty} \frac{(P + p_k + 1) - (P + 1)}{\log(P + 1)}$$
$$\geq \limsup_{k \to \infty} \frac{p_k}{\log P} \cdot \frac{\log P}{\log(P + 1)}$$
$$\geq \limsup_{k \to \infty} \frac{p_k}{(p_k - 1)\log 4} = \frac{1}{\log 4}$$

Bertrand's postulate implies

$$1 \leq \lim_{p \to \infty} \frac{\log p}{\log p_-} \leq \lim_{p \to \infty} \frac{\log 2p_-}{\log p_-} = 1$$

hence the corollary is proved. □

2.2.8 How Many Owl-Primes Are There?

The values of the polynomial $n^2 - n + 41$ for $n = 1, 2, 3, \ldots 40$ all turn out to be primes. This curious observation goes back to Euler. He was enough fluent with numerical calculations to get such an astonishing result. We now address the question which other primes besides 41 have the curious property noted by Euler.

Problem 22. *Find all primes up to 50 that have a similar property. In other words, find the primes $p \leq 50$ such that the list*

(2.2.26) $$n^2 - n + p \quad \text{for } 1 \leq n \leq p-1$$

consists only of primes.

Definition 9 (Owl-prime). I call a prime p an *owl-prime* [4] if the values of the polynomial $n^2 - n + p$ for $n = 1, 2, 3, \ldots p-1$ all turn out to be primes.

At first to begin such a calculation by hand is possible. I begin with $p = 2, 3, 5$.

Problem 23. *As a warm-up, find the three smallest owl-primes.*

Answer. It is easy to check that $p = 2, 3, 5$ are indeed owl-primes.

But for larger primes, it begins harder and harder to just use trial and error. So my next step is to cut down the list of possible owl-primes.

Proposition 23. *For any prime $p > 5$ equivalent are*

(a) *For any natural number n, the values $n^2 - n + p$ are divisible neither by $2, 3$ nor 5.*

(b) *p is congruent to either 11 or 17 modulo 30.*

Remark. Note this proposition gives only a *necessary* condition for a prime being an owl-prime.

Reason. Clearly $n^2 - n + p$ is odd since $n^2 - n$ is always even, and every prime $p > 2$ is odd.

If we go through the natural numbers $n = 1, 2, 3, \ldots$, the value of $n^2 - n$ is either divisible by three and thus congruent to 0 modulo 3, or it is congruent to 2 modulo 3. The first case occurs for $n \equiv 0$ or $n \equiv 1$ modulo 3. The second case occurs for $n \equiv 2 \mod 3$. To assure that $n^2 - n + p$ is never divisible by 3 for any n, we need to have $p \equiv 2 \mod 3$. Indeed in that case holds either $n^2 - n + p \equiv 2 \mod 3$ or $n^2 - n + p \equiv 2 + 2 \equiv 1 \mod 3$.

From the following small table, we get the possible values of $n^2 - n$ mod 5:

---

[4] I have chosen this nickname in honor of Euler

2.2. MORE ABOUT PRIMES

n	$n^2 - n$	$n^2 - n \mod 5$
1	0	0
2	2	2
3	6	1
4	12	2
5	20	0

We see $0, 1, 2$ are the possible values of $n^2 - n \mod 5$. To avoid that $n^2 - n + p$ is divisible by 5, we need that the sum $(n^2 - n) + p$ is different from zero modulo 5. To this end, $p \mod 5$ has to be different from 3 or 4. Moreover p is a prime and larger than 5. Hence the number p has to be congruent to either 1 or 2 modulo 5.

Problem 24. *If a number p is odd, $p \equiv 2 \mod 3$, and $p \equiv 1 \mod 5$, determine the number p modulo 30.*

Similarly, assume p is odd, $p \equiv 2 \mod 3$, and $p \equiv 2 \mod 5$, and determine p modulo 30.

Answer. In the first case, we get $p \equiv 11 \mod 30$. In the second case, we get $p \equiv 17 \mod 30$.

□

Now we have reduced the obvious question about further owl-primes between 5 and 50 considerably. As possible owl-primes between 5 and 50, there are only left the four numbers $11, 17, 41, 47$.

Problem 25. *Determine which ones of these four numbers $11, 17, 41, 47$ are really owl-primes.*

Before doing a brute-force calculation, let us cut down the amount of computation even further. It is well known that for the odd number N to be prime, it is sufficient that N has no odd divisors q for which $q^2 \leq N$ except 1. It is even enough to check the cases were q is a prime number. The function $n \mapsto n^2 - n$ is increasing for $n > \frac{1}{2}$. Since $n < p$ we see that all values N in the list (2.2.26) are less than $p^2 - p + p = p^2$. Hence the divisors q for the composite cases satisfy $q^2 \leq N < p^2$ and hence are all strictly less than p. For any prime $p > 2$, clearly $n^2 - n + p$ is odd since $n(n-1)$ is even and p is odd. Hence to have an owl-prime, it is sufficient to check that the numbers of the list (2.2.26) have no odd prime divisors less than p.

One may cut down the calculation still a bid more and get the following

Lemma 11. *The odd prime number p is an owl-prime if and only if*

> *q is not divisor of $n^2 - n + p$ for all odd primes $q < p$ and all $2 \leq n \leq \frac{q+1}{2}$*

Reason. To check the odd divisors q of $N = n^2 - n + p$, we consider the remainder values $r = \text{Mod}\,[n^2 - n, q]$. I use remainders in the range $0 \leq r \leq q - 1$ and denote by $\text{Mod}\,[N, q]$ the remainder for the number N in the division by q. Recall that the number $n^2 - n + p$ is odd. The number $n^2 - n + p$ is composite iff it has an odd divisor $1 < q < p$, which happens if and only if $\text{Mod}\,[n^2 - n + p, q] \neq 0$.

Take the contrary case. The number $n^2 - n + p$ from the list (2.2.26) is prime if and only if

$$\text{Mod}\,[n^2 - n + p, q] \neq 0 \quad \text{for all odd primes } q < p$$

The number p is an owl-prime if and only if

$$\text{Mod}\,[n^2 - n + p, q] \neq 0 \quad \text{for all odd primes } q < p$$
$$\text{and all } 1 \leq n \leq \tfrac{q+1}{2}$$

One needs still to explain why the range for n is not the entire interval $1 \leq n \leq p - 1$. The restriction $n < q$ is possible since $\text{Mod}\,[n^2 - n, q]$ depends only of the remainder $\text{Mod}\,[n, q]$. The stronger restriction $n \leq (q+1)/2$ is possible because of

$$\text{Mod}\,[n^2 - n, q] = \text{Mod}\,[m^2 - m, q] \quad \text{for } m = q + 1 - n$$

Finally, the case $n = 1$ can be skipped because of the assumption that p is an odd prime. □

Let me take the example $p = 17$. Is this an owl-prime? The divisors 3 and 5 have been excluded earlier. One needs still to check the cases

$$q = 7 \text{ and } 2 \leq n \leq 4$$
$$q = 11 \text{ and } 2 \leq n \leq 6$$
$$q = 13 \text{ and } 2 \leq n \leq 7$$

You see that even with all the reductions done above, some computation is still left to be done. In the above instance, these are 27 cases to be checked.

After some computations, one sees that 11 and 17 are indeed owl-primes. To get Euler's owl-prime 41, the computations are still

2.2. MORE ABOUT PRIMES

a bid heavier. There are 187 cases to be checked. I wonder whether Euler had still another idea to reduce that nuisance of computation. Honestly, my best guess is, he has just worked it out.

To see that the other open case 47 is not an owl-prime is easier to see. With $n = 2$ one gets $n^2 - n + p = 49$ which is composite.

Open Problem. *Show that* $2, 3, 5, 11, 17$ *and* 41 *are the only owl-primes, or find a larger owl-prime.*

Corollary 15. *The odd prime number $p > 5$ is an owl-prime if and only if*

- $p \equiv 11 \mod 30$ *or* $p \equiv 17 \mod 30$
- $p + 2$ *is a prim, too;*
- $n^2 - n + p$ *is a prim for* $3 \leq n \leq \frac{p-3}{2}$

Explanation. We have shown that these three items hold for the owl-primes. Note that the second item is the special case of the third with $n = 2$. In other words, the three items are necessary.

Conversely, the three items are sufficient conditions for an owl-prime. Indeed since p and $p + 2$ are primes, $p - 2$ is divisible by three and since $p > 5$ it is not a prime. Hence $q \leq p - 4$. Now the requirements from lemma 11 are satisfied, and hence p is an owl-prime. □

By means of this corollary, one can check with mathematica, within a minute of computation that there are no further owl-primes less that $3 \cdot 10^8$.

```
For[i = 0, i < 10^7, i++,
   p = 11 + 30 i; good = PrimeQ[p] && PrimeQ[p + 2];
   n = 3; While[good && n <=  (p - 3)/2 ,
     good = good && PrimeQ[n^2 - n + p]; n++];
   If[good, Print[p]]];
```

11

41

```
For[i = 0, i < 10^7, i++,
   p = 17 + 30 i; good = PrimeQ[p] && PrimeQ[p + 2];
   n = 3; While[good && n <=  (p - 3)/2 ,
     good = good && PrimeQ[n^2 - n + p]; n++];
```

```
If[good, Print[p]]];
```

17

Here is an additional observation.

Remark. Let q be an odd prime. The remainders $\text{Mod}\,[n^2 - n, q]$ are all different for $1 \leq n \leq (q+1)/2$, but the same remainders occur for n and $q+1-n$. Hence one gets with the above values already all remainders of the set

$$L(q) = \{\text{Mod}\,[n^2 - n, q] : 1 \leq n \leq q\}$$

Each remainder occurs exactly twice, except the one for $n = (q+1)/2$.

Reason. We use the formula $n(n-1) - m(m-1) = (n-m)(n+m-1)$. Hence one gets the equivalence

$$\text{Mod}\,[n^2 - 2, q] = \text{Mod}\,[m^2 - m, q] \Leftrightarrow \text{Mod}\,[(n-m)(n+m-1), q] = 0$$

Since q is assumed to be a prime one gets even more:

$$\text{Mod}\,[n^2 - 2, q] = \text{Mod}\,[m^2 - m, q]$$
$$\Leftrightarrow \text{either } m \equiv n \mod q \text{ or } m \equiv q+1-n \mod q$$

The remainders in the set $L(q)$ are already covered by

$$L[q] = \{\text{Mod}\,[n^2 - n, q] : 1 \leq n \leq (q+1)/2\}$$

For $(q+1)/2 \leq n \leq q$ the same values are repeated in reversed order, except the middle value for $n = (q+1)/2$ that appears only once. □

Take the examples $q = 5$, $q = 7$ and $q = 11$. One gets in that order

$$L(5) = [0, 2, 1, 2, 0]$$
$$L(7) = [0, 2, 6, 5, 6, 2, 0]$$
$$L(11) = [0, 2, 6, 1, 9, 8, 9, 1, 6, 2, 0]$$

We have a found nice way to produce palindromes.

2.2. MORE ABOUT PRIMES

2.2.9 Arithmetic Progressions of Primes

This is a famous playground for computations of huge primes. Therefore it is worth while to look at wikipedia:

https://en.wikipedia.org/wiki/Primes_in_arithmetic_progression

https://nrich.maths.org/6413/solution

Definition 10 (Arithmetic progression). A finite sequence

(2.2.27) $\qquad a, a+d, a+2d, \ldots, a+(n-1)d$

is called an *arithmetic progression*. The leading term is a, the difference is d, and the length is n. We shall always assume in the following $d \geq 2, n \geq 2$, but the first term a may be any integer.

An arithmetic progression that consists of only primes and length $n \geq 2$ and $d \geq 2$ is called an AP-n.

Here are a few questions about these AP-n, of very different degrees of difficulty:

(i) Does there exist an AP-2 of length three.

(ii) Do there exist infinitely many AP-2 of length two. These are also called *twin primes*.

(iii) Does there exists only one AP-3 with difference $d = 4$, or several ones.

(iv) Does there exists only one AP-5 with difference $d = 6$, or several ones.

(v) This unique one AP-5 with difference $d = 6$ is easy to find with calculation by heart, Find it without use of pen and paper!

(vi) Do there exist AP-n of arbitrary large n.

(vii) Is there an upper bound for the length n, in terms of the leading term a.

(viii) Find several AP-4 with difference 6. These are called *sexy prime quadruplets*.

(ix) What is the maximal length of any AP-n with difference 10.

(x) Do there exist infinitely many sexy prime quadruplets?

Next I derive some necessary restrictions for the leading term, difference and length of any arithmetic progression of primes.

Lemma 12. *Assume $2 \leq m \leq n$ and $\gcd(m,d) = 1$. The remainders of the arithmetic sequence*

$$r_i = \operatorname{Mod}[a + id, m] \text{ for } i = 0, 1, \ldots, n-1$$

assume each value $0, 1, \ldots, m-1$ at least $\lfloor \frac{n}{m} \rfloor$ times and at most $\lceil \frac{n}{m} \rceil$ times. Especially, at least $\lfloor \frac{n}{m} \rfloor$ terms and at most $\lceil \frac{n}{m} \rceil$ terms are divisible by m.

Especially, if $p \leq n$ and p is a prime not dividing the difference d, then at least one term $a + id$ is divisible by p.

Proof. By Euclid's Lemma $r_i \equiv r_j \pmod{m}$ if and only if m divides $i - j$. The remainders $r_0, r_1, r_2, \ldots r_{m-1}$ are all different. For the indices $i \geq m$, these remainders are repeated periodically. Hence at least $\lfloor \frac{n}{m} \rfloor$ terms and at most $\lceil \frac{n}{m} \rceil$ terms of the arithmetic sequence (2.2.27) are divisible by m. □

Remark. In the case $m > n$ and $\gcd(d, m) = 1$, even all remainders $r_0, r_1, r_2, \ldots r_{n-1}$ are different, but not all possible remainders do occur.

We can now get some nice necessary conditions that any arithmetic sequence of primes satisfies.

Proposition 24 (Necessary properties of arithmetic progressions of primes). *For the arithmetic sequence*

$$a + id \quad \text{with } 0 \leq i \leq n-1$$

to consist only of primes it is necessary, but not sufficient, that either one of the following cases (i) *or* (ii) *occurs.*

(i) $n = a$ *and the difference d is divisible by the product of all primes <u>strictly less</u> than the number n of terms; or*

(ii) a *is a prime larger than n and the difference d is divisible by the product of all primes <u>less or equal</u> to the number n of terms.*

Definition 11 (primondial). The any natural number n, the *primondial* is defined as the product of all primes less or equal to n. The primondial of n is denoted by $n\#$.

2.2. MORE ABOUT PRIMES 63

If n is prime then an AP-n can begin with n and have a common difference which is only a multiple of $(n-1)\sharp$ instead of $n\sharp$. For example, the AP-3 with primes $3, 5, 7$ and common difference $d = 2\sharp = 2$, or the AP-5 with primes $5, 11, 17, 23, 29$ and common difference $d = 4\sharp = 6$.

Before digging into the proof of our basic proposition 24, a few remarks about the introductory problems are in place.

(i) Does there exists an AP-2 of length three. There is the example $3, 5, 7$. Here we have $n = a = 3$, the case (i) in the above proposition.

(ii) Do there exist infinitely many AP-2 of length two. These are also called *twin primes*. This is a famous open problem.

(iii) Does there exists only one AP-3 with difference $d = 4$, or several ones. With $n = 3$ and $d = 4$, the product of the primes less or equal n is 6, which does not divide d. But the product of all primes strictly less than n equals 2, and this divides d. So we are restricted to case (i) in the proposition above. The AP-3 has to start with $a = n = 3$ is turns out to be $3, 7, 11$. This is the only possible case.

(iv) Does there exists only one AP-5 with difference $d = 6$, or several ones. With $n = 5$ and $d = 6$, holds $n\sharp = 30$ for the product of the primes less or equal n. This is not a divisor of the difference $d = 6$. Only $(n-1)\sharp = 6$ is a divisor of 6. So we are restricted to case (i) in the proposition above. The AP-5 has to start with $a = n = 5$ is turns out to be

$$5, 11, 17, 23, 29$$

This is the only possible case.

(vi) Do there exist AP-n of arbitrary large n. According to the celebrated Green–Tao theorem, there exist arbitrarily long sequences of primes in arithmetic progression. So the answer is known to be yes,—after considerable hard work.

(vii) Is there an upper bound for the length n, in terms of the leading term a. Yes, the length n has the upper bound $n \leq a$. We prove this below by the first easy lemma 13.

(viii) Find several AP-4 with difference 6. These are called *sexy prime quadruplets*. With $(n-1)\sharp = 6 = d$ we may use case

(ii) and get hopefully many possibilities for the leading term a. By the above remark 2.2.9 with $m = 5$, that the four sexy primes
$$a, a + 1 \cdot 6, a + 2 \cdot 6, a + 3 \cdot 6, a + 4 \cdot 6$$
are different remainders modulo 5. These remainder are all nonzero, except for the special case from item (iv) above. We get the remainders $1, 2, 3, 4$. They have to occur in this order. Since primes are odd, the last digits, which are the remainders modulo 10, have to be $1, 7, 3, 9$. Now one may just try the cases $11, 31, 41, 61, 71, 101, 131$ for the leading term a. After having ruled out the cases which still contain composite numbers, three cases for sexy prime quadruplets are left over

$$11, 17, 23, 29$$
$$41, 47, 53, 59$$
$$61, 67, 73, 79$$

(ix) What is the maximal length of any AP-n with difference 10. We have just obtained the example $31, 41$ and get already existence and a maximal length at least 2. Since the numbers 3 and 10 are relatively prime, we may use lemma 12. Exactly one of the three numbers of any arithmetic sequence of difference 10 is divisible by 3. We see that $n = 2$ is indeed the maximal length.

(x) Do there exist infinitely many sexy primes quadruplets? There is an ongoing computer chase for some very large such quadruplets.

https://en.wikipedia.org/wiki/Sexy_prime

It follows from widely believed conjectures, such as Dickson's conjecture and some variants of the prime k-tuple conjecture, that if $p > 2$ is the smallest prime not dividing d, then there are infinitely many AP-(p-1) with common difference d. For example, 5 is the smallest prime not dividing 6, so there is expected to be infinitely many AP-4 with common difference 6, which is called a sexy prime quadruplet. When $a = 2, p = 3$, it is the twin prime conjecture, with an "AP-2" of 2 primes $(a, a + 2)$.

2.2. MORE ABOUT PRIMES

(xi) Does for all primes n exist an AP-n with the leading term $a = n$. This is a conjecture believed to be true. One may read on wikipedia that it is conjectured that examples for case (i) with leading term $n = a$ and length n exist for all primes n. But one has to allows the difference d to be a large multiple of $(n-1)\sharp$. As of 2018, the largest prime for which this is confirmed is $n = 19$. The AP-19 was found by Wojciech Izkowski in 2013:

$$19 + 4244193265542951705 \cdot 17\sharp \text{ for } 0 \leq n \leq 18$$

The numbers get huge, indeed.

Problem 26. *Does there exist any more AP-n with $a = n$ and $d = (n-1)\sharp$, except for the two cases with $n = 3$ and $n = 5$. This may be a difficult question.*

Problem 27. *Is it true or not? Let q be an odd prime. There exists at most one AP-q with difference $d = q + 1$.*

Answer. For the notation see also definition 11. With $n = q$ and $d = q + 1$, the stronger condition $n\sharp \mid d$ does not hold. The case (ii) is excluded. We might still have case (i) with $n = q = a$, and hence one gets at most one solution.

Problem 28. *Cut down the cases for which there really exists one AP-q with difference $d = q + 1$.*

Answer. One still needs the weaker assumption $(n-1)\sharp \mid d$. In the present case with $n = q$ and $d = q+1$, this means that the product of all odd primes $r \leq q-1$ has to divide $q+1$. This works only for $q = 3$ and $q = 5$, but not for any lager prime; the product on the lower side grows too fast for larger q. One gets a positive answer only in the above two cases $q = 3$ and $q = 5$.

Problem 29. *A sexy primes quadruplet can either begin with a prime congruent to 1 modulo 6, or a prime congruent to 5 modulo 6. Do both cases occur?*

Answer. Below $1\,000$ there are the quadruplets

$$(5, 11, 17, 23, 29), (41, 47, 53, 59), (61, 67, 73, 79),$$
$$(251, 257, 263, 269), (601, 607, 613, 619), (641, 647, 653, 659)$$

After the beginning exceptional AP-5, the two cases alternate. Both cases are possible.

The five largest known sexy prime quadruplets, announced 2005 through October 2019 in wikipedia, turn all out to be equivalent to 1 modulo 6. They have between 1002 and 3025 digits. They can still easily checked to be correct by mathematica. But what significance has the first observation?

The proof of proposition 24 begins with three lemmas.

Lemma 13. *For any AP-n holds* $\gcd(a,d) = 1$ *and* $n \leq a$.

Reason. In the case $g = \gcd(a,d) > 1$, the second term $a+d$ would have the divisor g different from one and itself, and cannot be a prime number. The $a+1$-th number in the continued sequence is $a + a \cdot d = a(1+a)$, which is composite. Since this number cannot be part of the AP-n, we get $n \leq a$. □

Lemma 14. *For any AP-n with* $n = a$, *the difference d is divisible by the product of all primes strictly less than the number n of terms.*

Proof. Assume towards a contradiction that $n = a$ and the difference d is not divisible by the product of all primes strictly less than the number n of terms. Let $p < n$ be a prime which does not divide d.

By the last part of lemma 12, at least one term $a + id$ in the progression is divisible by p. The remainders $\text{Mod}\,[a + id, p]$ for $0 \leq i \leq n-1$ take all p possible values, and one of the remainders gets zero. Since the latter term $a + id$ is assumed to be a prime number, one concludes $p = a + id$. In the end one gets the inequalities

$$a \leq a + id = p < n = a$$

which is impossible. □

Lemma 15. *For any AP-n with* $n < a$, *the difference d is divisible by the product of all primes less or equal than the number n of terms.*

Proof. The proof is *almost* as above, but I prefer to get a bid lengthy and spell out the necessary modifications. Assume towards a contradiction that $n < a$ and the difference d is not divisible by the product of all primes *less or equal* to the number n of terms. Let $p \leq n$ be a prime which does not divide d. By the last part of lemma 12, at least one term $a + id$ in the progression is divisible by p. Since this term $a + id$ is assumed to be a prime holds $p = a + id$. In the end one gets the inequalities

$$a \leq a + i \cdot d = p \leq n < a$$

2.2. MORE ABOUT PRIMES 67

which is impossible. Note that in comparison to the previous lemma, the places for equality and strict inequality in the last line have been switched. □

End of the proof for proposition 24. In case (i) one would get $a \leq p < n = a$, in case (ii) one would get $a \leq p \leq n < a$ for the tentative prime p,—hence this prime does not exist. □

With the notation for the primondial 11, the assumption of proposition 24 are

$$\text{case (i):} \quad a = n \text{ and } (n-1)\sharp \mid d$$
$$\text{case (ii):} \quad a < n \text{ and } n\sharp \mid d$$

Remark. For these necessary conditions, one needs neither assume that the length n is a prime, nor that the progression is extended maximally.

The following corollary is just another way to express the same result.

Corollary 16. *Take any AP-n and let p be the least prime that does not divide the difference d.*

The number of terms $n < p$ is either less than prime p, or there occurs the special case $n = p$, in which the prime $p = a$ is the first term of the sequence.

Proof. In the case (i) of the main proposition 24 hold $n = a$ and $(n-1)\sharp \mid d$. Hence all primes less than n are divisors of d. The next prime $p = n$ cannot be a divisor of d since $\gcd(a, d) = 1$.

In the case (ii) holds $n < a$ and $n\sharp \mid d$. Hence all primes less or equal to n are divisors of d. The least prime that is not a divisor of d is larger than n. □

Definition 12. The AP-n is called to be *extended maximally* or simply *maximal* iff its is not a subset of any longer progression of only primes. Equivalently, one has to satisfy the two assumptions

- $a + n \cdot d$ is a composite number;
- either $a - d$ is a composite number or $a \leq d + 1$.

It is tempting to guess that the maximality gives easy additional information about the parameters a, n and d. But I had only very little success.

Problem 30. *Prove or disprove: If $q \nmid d$ is a prime different to the leading term of the given AP-n, then $q > n$. We even get that $q > n^+$ is larger than the length of the AP-n^+ extended forward.*

Answer. The assertion is true. In the case $n < a$ holds $n\sharp \mid d$ by our proposition 24. Hence all primes less or equal to n divide the difference d. By the contrapositive, we see that any prime not dividing the difference d has to be greater than n.

In the case $a = n$ holds $(n-1)\sharp \mid d$ by our proposition 24. Hence all primes less to n divide the difference d. By the contrapositive, we see that any prime not dividing the difference d has to be greater or equal to n. We have to exclude $q = n$. But in that case holds $q = n = a$, and $q = a$ has been excluded.

Problem 31. *Prove that any AP-n with leading term $a = n$ cannot be extended.*

Answer. For all AP-n with $a = n$ already holds $a \leq d+1$, which makes a backward extension impossible. Indeed by case (i) from proposition 24 is obtained

$$a = n \leq (n-1)\sharp + 1 \leq d+1$$

Hence maximality is equivalent to $a + n \cdot d$ being composite, which indeed holds because of $a + n \cdot d = a(1 + d)$.

Problem 32. *Take the case of an AP-n with $a > n$ and $a - d > 1$ which cannot be extended backwards. Get a lower bound for the leading term a.*

Answer. The assumptions mean that $a - d$ is a composite number. Let q be the least prime factor of the composite number $a - d$. Hence we get from proposition that $n\sharp \mid d$ and hence

$$q^2 \leq a - d \leq a - n\sharp$$

Moreover $\gcd(a - d, d) = \gcd(a, d) = 1$. Hence $n\sharp \mid d$ implies that all prime factors $q \mid a - d$ satisfy $q > n$. Hence

$$(n+1)^2 + n\sharp \leq q^2 + n\sharp \leq a$$

Problem 33. *What can one conclude for $n = 2$, what for $n = 3$?*

Answer. With $n = 2$ one gets that $11 \leq a$ holds for all twin primes except the leading $3, 5, 7$. Note that neither the pair $3, 5$ nor the pair $5, 7$ satisfies the assumptions made in problem 32.

With $n = 3$ one gets for the sexy primes $24 = 16 + 6 \leq a$, with the exception of the leading $5, 11, 17, 23, 29$. The next smallest sexy primes are $41, 47, 53, 59$.

Chapter 3

Early Achievements

3.1 The Chinese Remainder Theorem

3.1.1 Simultaneous Congruences

Problem 34. *Given that*

$$x \equiv 19 \mod 765 \quad \text{and} \quad x \equiv 1 \mod 567$$

Find the smallest positive solution for x.

Answer. One can determine the unknown integer x modulo the least common multiple of lcm(765, 567). To get a solution, one needs p and q such that $x = 19 + p \cdot 765 = 1 + q \cdot 567$. Since $\frac{19-1}{9} = 2$ is an integer, the congruence is solvable. We need to multiply the result of the problem above by this integer 2 and get

$$9 = (-20) \cdot 765 + 27 \cdot 567$$
$$18 = (-40) \cdot 765 + 54 \cdot 567$$
$$18 + 40 \cdot 765 = 54 \cdot 567$$
$$19 + 40 \cdot 765 = 1 + 54 \cdot 567$$
$$30\,619 = 30\,619$$

We get a solution $x = 30\,619$, which turns out to be the smallest one, in this example. [1]

---

[1] One does not always get immediately the smallest solution. The values p and q are not needed any more.

Problem 35. *Solve the simultaneous congruences*

(3.1.1)
$$a \equiv 2 \pmod{121}$$
$$a \equiv 5 \pmod{23}$$

Answer. We want to calculate with a sequence of congruences modulo the least common multiple $m = \text{lcm}[121, 23] = 121 \cdot 23 = 2783$, and use the same operations as done by the Euclidean algorithm to get the greatest common divisor of 121 and 23.

q_i	s_i	r_i	congruence	
	1	121	$121(a-2)$	$\equiv 121(5-2)$
5	0	23	$23(a-2)$	$\equiv 0$
3	1	6	$6(a-2)$	$\equiv 363$
1	3	5	$5(a-2)$	$\equiv (-3) \cdot 363$
5	4	1	$1(a-2)$	$\equiv 4 \cdot 363$
0	23	0	$0(a-2)$	$\equiv (-23) \cdot 363$

One gets $a \equiv 2 + 4 \cdot 363 = 1454 \pmod{2783}$, which can easily be checked to be a solution of the system (3.1.1).

Problem 36. *Find integers s and t such that $22s + 8t = 2$. Find the least common multiple of 8 and 22.*

Answer. One can get immediately $\gcd(22, 8) = 2$. Hence the least common multiple of 22 and 8 is 88. The extended Euclidean algorithm gives the greatest common divisor as linear combination:

row 0:   1   0
row 1: $22 : 8 = 2$ rem 6   0   1
row 2: $8 : 6 = 1$ rem 2   1   2
row 3: $6 : 2 = 3$ rem 0   1   3

Indeed, $\gcd(22, 8) = 2 = (-1) \cdot 22 + 3 \cdot 8$.

Problem 37. *Given that*

$$x \equiv 3 \mod 8 \quad \text{and} \quad x \equiv 5 \mod 22$$

determine the unknown integer x modulo the least common multiple of 8 and 22.

3.1. THE CHINESE REMAINDER THEOREM

Answer. One needs s and t such that $3 + 8s = 5 + 22t$. This works for $3 + 8 \cdot 3 = 5 + 22 \cdot 1 = 27$. Hence both

$$x \equiv 27 \mod 8 \quad \text{and} \quad x \equiv 27 \mod 22$$
$$\text{what implies} \quad x \equiv 27 \mod 88$$

We now turn to the general case of simultaneous congruences

$$(3.1.2) \qquad \begin{aligned} x &\equiv u \pmod{a} \\ x &\equiv v \pmod{b} \end{aligned}$$

We want to calculate with a sequence of congruences modulo $m = ab$, and use the same sequence of operations as those occurring in the Euclidean algorithm determining the greatest common divisor $\gcd(a, b)$. For an appropriate new start, the system (ch) is changed to the equivalent equations

$$(3.1.3) \qquad \begin{aligned} a(x - u) &\equiv a(v - u) \pmod{ab} \\ b(x - u) &\equiv 0 \pmod{ab} \end{aligned}$$

Note that the second equation of system (ch) has been multiplied with a, and the first equation has been multiplied with by b, moreover the equations have been switched. The extended Euclidean algorithm calculates the successive remainders r_i and quotients q_i via

$$(1.1.7) \quad r_0 = a, \ r_1 = b, \ r_{i+1} = r_{i-1} - q_i r_i \ \text{and} \ 0 \le r_{i+1} < r_i$$

and gets the sequences s_i t_i by the recursions

$$(3.1.4) \qquad \begin{aligned} s_0 &= 1, \ s_1 = 0, \ s_{i+1} = s_{i-1} + q_i s_i \\ t_0 &= 0, \ t_1 = 1, \ t_{i+1} = t_{i-1} + q_i t_i \end{aligned}$$

for $i = 1, 2, \ldots, m$. The algorithm stops when a zero remainder $r_{m+1} = 0$ appears for the first time in row m. The last nonzero remainder is the greatest common divisor.

$$(3.1.5) \qquad r_m = \gcd(a, b), \ r_{m+1} = 0$$

We want to calculate with a sequence of congruences modulo the least common multiple $m = ab$. As already shown in the example, we use the system (3.1.3) in rows with $i = 0$ and $i = 1$ and perform the same sequence of operations as in Euclidean algorithm. In this manner, we get a sequence of congruences

$$(3.1.6) \qquad r_i(x - u) \equiv a(v - u) \cdot (-1)^i s_i \pmod{ab}$$

for $i = 0, 1, 2, \ldots, m+1$. From row $i = m$ we conclude, after division by $r_m = \gcd(a,b)$,

(3.1.7) $\qquad x \equiv u + a(v-u) \cdot (-1)^m s_m \pmod{\operatorname{lcm}[a,b]}$

We have shown earlier that the recursions (3.1.4) imply

$$s_{m+1} = \frac{b}{\gcd(a,b)}, \quad t_{m+1} = \frac{a}{\gcd(a,b)}$$

Hence the equation from row $i = m+1$ tells that $0 \equiv a(v-u) \cdot s_{m+1} \pmod{ab}$ and hence

(3.1.8) $\qquad 0 \equiv v - u \pmod{\gcd(a,b)}$

It is easy to convince oneself that any two successive rows obtained during the Euclidean algorithm are equivalent to the first and second row. Hence the original system (ch) is equivalent to the two congruences (3.1.7) together with (3.1.8). In other words, we have proved

Proposition 25. *The solution of the simultaneous congruences (ch) can be calculated by the extended Euclidean algorithm. One obtains the result (3.1.7). But this is a valid solution if and only if the solvability condition (3.1.8) holds.*

Problem 38. *Convince yourself that in case the solvability condition (3.1.8) not hold, the result (3.1.7) is at least a solution of the relaxed system*

(3.1.9)
$$x \equiv u \mod \frac{a}{\gcd(a,b)}$$
$$x \equiv v \mod \frac{b}{\gcd(a,b)}$$

Simultaneous congruences program for the TI84

The program asks for the numbers u, a, v, b from the simultaneous congruences

(ch)
$$x \equiv u \mod a$$
$$x \equiv v \mod b$$

Three successive remainders r_i are

X,Y,Z

3.1. THE CHINESE REMAINDER THEOREM

Three successive terms $(-1)^i s_i$ are

R,S,G

In the end, one should have
$r_M = Y$, $(-1)^M s_M = S$, $(-1)^{M+1} s_{M+1} = G$.

Unfortunately, in the version written down, I needed to replace list L_1 by L1, and $\neq$ by not= . Please find out from the context.

PROGRAM:TRY

```
Prompt U,A
Prompt V,B
A -> Y: B -> Z
1 -> S: 0 -> G: 0 -> M
Lbl 1
If Z = 0:Goto 2
Y -> X: Z -> Y
S -> R: G -> S
int(X/Y) -> Q
X-QY -> Z: R-QS -> G: M+1 -> M
Goto 1
Lbl 2
Disp{Y,M}
B/Y -> H: AH -> L
(U-V)/Y -> D
DS -> P
P-int(P/H)H -> P
U-PA -> W
W-int(W/L)L -> W
Lbl 3
{D,W-U-int((W-U)/A)A,W-V-int((W-V)/B)B}
    -> L_1
Disp L_1
Disp {W,L}
```

The output of the program consists of these three lists:

{Y,M}

{D,W-U-int((W-U)/A)A,W-V-int((W-V)/B)B}

if and only if the congruences are solvable, and two check numbers that are 0,0 if and only if the congruences solved correctly;

{W,L}

which yields the actual solution $x \equiv w \mod l$.

3.1.2 Chinese Remainder (Sun-Ze's) Theorem

At this point, we additionally assume the integers a and b to be <u>relatively prime</u>. The solution of the simultaneous congruences (ch) defines a (total) function $f : \mathbf{Z}_a \times \mathbf{Z}_b \mapsto \mathbf{Z}_{ab}$ since we know the simultaneous congruences to be solvable for any right-hand side $(u,v) \in \mathbf{Z}_a \times \mathbf{Z}_b$, and the solution to be uniquely determined modulo ab.

The inverse function $x \mapsto f^{-1}(x) = (u,v)$ is obtained by reducing any $x \in \mathbf{Z}_{ab}$ to the remainders $u \in \mathbf{Z}_a$ and $v \in \mathbf{Z}_b$; as one naturally has to do when checking the solution of the Chinese remainder problem. Since the reduced remainders are unique, we see that f^{-1} is a function, too, and hence the function f is injective. Moreover, the value $x \in \mathbf{Z}_{ab}$ can be chosen arbitrarily. Hence f^{-1} is a well-defined (total) function, and hence f is surjective.

Sun's-Ze's result was obtained before 850, the statement below was obtained by Chin-Chin Shao about 1250.

Theorem 8 (Chinese Remainder Theorem). *Assume a and b are relatively prime positive integers. Let s,t be integers satisfying*

(3.1.10) $$\gcd(a,b) = sa - tb$$

obtained via the extended Euclidean algorithm. The function

$$f : \mathbf{Z}_a \times \mathbf{Z}_b \mapsto \mathbf{Z}_{ab}$$
$$f(u,v) = v \cdot sa - u \cdot tb$$

is a bijection which yields the solution of the Chinese remainder problem

(ch) $$\begin{aligned} x &\equiv u \mod a \\ x &\equiv v \mod b \end{aligned}$$

Its inverse $x \mapsto f^{-1}(x) = (u,v)$ is obtained by reducing $x \in \mathbf{Z}_{ab}$ to the remainders $u \in \mathbf{Z}_a$ and $v \in \mathbf{Z}_b$.

3.2 The Geometric Series

We put $0^0 = 1$ in this book, unless stated otherwise. We use the convention from mathematica that the greatest common divisor is always positive, except for $\gcd(0,0) = 0$.

Problem 39. *The sum formula for the finite geometric series:*

$$(3.2.1) \qquad \sum_{k=0}^{n-1} x^k = \frac{1-x^n}{1-x}$$

holds for all $x \neq 1$ and $n \geq 1$. Prove the formula by induction.

Remark. In the case $x = 0$, one interprets the sum as $1 + 0 + \ldots$ with $0^0 = 1$ and $0^k = 0$ for $k \geq 1$. In the same way are treated all power series.

Answer. Induction start $n = 1$:

$$\text{l.h.s.} = x^0 = 1$$
$$\text{r.h.s.} = \frac{1-x}{1-x} = 1$$

since $x \neq 1$. Thus the left-hand side turns out to be equal to the right-hand side. The case $x = 0$ is included by convention.
Induction step "$n \mapsto n+1$":

Begin with the left-hand side for the formula with n replaced everywhere by $n+1$:

$$\text{l.h.s.} = \sum_{k=0}^{n} x^k = \sum_{k=0}^{n-1} x^k + x^n \qquad \text{by recursive definition of the sum}$$
$$= \frac{1-x^n}{1-x} + x^n \qquad \text{by induction assumption}$$
$$= \frac{1-x^n + x^n(1-x)}{1-x} = \frac{1-x^{n+1}}{1-x} \qquad \text{by calculation}$$
$$= \text{r.h.s.} \qquad \text{is equal to the right-hand side.}$$

Hence we have shown the sum formula holds for $n+1$. By induction, it holds for all $n \in \mathbf{N}$.

3.2.1 My Little Theorems about Powers

Though this subsection, a, b, n, m, q are integers and $n, m \geq 0, q \geq 1$.

Problem 40.

(3.2.2) $a^n - 1 \mid a^{qn} - 1$ for all $q \geq 1, n \geq 0$

(3.2.3) $a^n + 1 \mid a^{qn} + 1$ for all $q \geq 1$ odd and $n \geq 0$

Answer. The cases with $a = 0$ or $n = 0$ can ge checked directly. Assume now $a \neq 0$ and $n \geq 1$. From the geometric sum (3.2.1) with $x := a^q$ or $x := -(a^q)$ one gets

$$a^{qn} - 1 = (a^q - 1)\sum_{k=0}^{n-1} a^{qk} \text{ and } (-a^q)^n - 1 = (-a^q - 1)\sum_{k=0}^{n-1} (-a^q)^k$$

confirming the claims.

Proposition 26 (The little proposition). *I use the convention from mathematica that the greatest common divisor is always positive, except for* $\gcd(0,0) = 0$. *Given are integers* $a \neq 0$ *and* $n, m, k, l \geq 0$. *If k divides l, then $a^k - 1$ divides $a^l - 1$. Moreover*

(3.2.4)
$\gcd(a^m - 1, a^n - 1) = |a^{\gcd(m,n)} - 1|$ *in all cases;*

(3.2.5)
$\gcd(a^m - 1, a^n - 1) = a^{\gcd(m,n)} - 1$ *if $a > 0$ or n, m both even;*

(3.2.6)
$\gcd(a^m - 1, a^n - 1) = -a^{\gcd(m,n)} + 1$ *if $a < 0$, and m or n is odd.*

Proof. The case with $a \neq 0$ and $mn = 0$ can be checked directly. Too, the case $a = 0$ and $m \geq 1$, $n \geq 1$ can be checked directly. Assume now $a \neq 0$ and $m, n \geq 1$. To check the first part, assume that $m = sn$. The geometric series with quotient $q := a^n$ yields

$$a^m - 1 = q^s - 1 = (q-1)(1 + q + q^2 + \cdots + q^{s-1})$$

Hence $a^m - 1$ is divisible by $a^n - 1 = q - 1$.

We see from this first step that $a^{\gcd(m,n)} - 1$ is a divisor of $\gcd(a^m - 1, a^n - 1)$. To confirm the reversed divisibility, the Euclidean algorithm is needed. We begin with the first division with remainder $m = qn + r$ with $0 \leq r < n$ and see

$$a^n - 1 \mid (a^{qn} - 1)a^r$$
$$\gcd(a^m - 1, a^n - 1) \mid \gcd(a^{qn+r} - 1, a^{qn+r} - a^r) \mid a^r - 1$$
$$\gcd(a^m - 1, a^n - 1) \mid \gcd(a^n - 1, a^r - 1)$$

3.2. THE GEOMETRIC SERIES

Let $r_0 = m, r_1 = n, r_2, \ldots, r_M = \gcd(m,n), r_{M+1} = 0$ be the sequence of remainders occurring in the Euclidean algorithm. Successively we get

$$\gcd(a^m - 1, a^n - 1) \mid \gcd(a^n - 1, a^r - 1)$$
$$\mid \gcd(a^{r_2} - 1, a^{r_3} - 1) \mid \cdots$$
$$\cdots \mid \gcd(a^{r_M} - 1, a^{r_{M+1}} - 1)$$
$$= \gcd(a^{\gcd(m,n)} - 1, 0) = a^{\gcd(m,n)} - 1$$

More easily we see that $a^{\gcd(m,n)} - 1$ is a divisor of both $a^m - 1$ and $a^n - 1$, and hence of $\gcd(a^m - 1, a^n - 1)$. We conclude the equality (3.2.4), as claimed. I use the convention from mathematica that the greatest common divisor is always positive, except for $\gcd(0,0) = 0$. Note that we still have the awkward absolute value $|a^{\gcd(m,n)} - 1|$. In the case $a < 0$ and $\gcd(m,n)$ odd, one gets $a^{\gcd(m,n)} - 1 \leq -2$, and hence $|a^{\gcd(m,n)} - 1| = -a^{\gcd(m,n)} + 1$. Hence the reversed sign in the third equation. □

Remark. The converse divisibility can be checked via the extended Euclidean algorithm, too. There exist natural numbers $s \geq 0$ and $t \geq 0$ such that $sm - tn = \gcd(m,n)$ (possibly after switching m with n). Assume that d divides both $a^m - 1$ and $a^n - 1$. By the first part of the Lemma, the number d divides both $a^{sm} - 1$ and $a^{tn} - 1$, and hence their difference

$$(a^{sm} - 1) - (a^{tn} - 1) = (a^{sm-tn} - 1)a^{tn} = (a^{\gcd(m,n)} - 1)a^{tn}$$

The base a and hence a^{tn} are relatively prime to $a^m - 1$ and hence to d. Hence d divides $a^{\gcd(m,n)} - 1$. Since this reasoning applies for any common divisor of $a^m - 1$ and $a^n - 1$, we conclude that $\gcd[a^m - 1, a^n - 1]$ is a divisor of $a^{\gcd(m,n)} - 1$.

Lemma 16. *Assume p is a prime, and the number $c \neq 1$ is such that $p \mid c - 1$. Then*

$$p \mid \frac{c^p - 1}{c - 1}$$

Proof. We use the geometric series and the congruence $c \equiv 1 \mod p$.

$$\frac{c^p - 1}{c - 1} = \sum_{j=0}^{p-1} c^j \equiv \sum_{j=0}^{p-1} 1 = p \equiv 0 \mod p$$

Hence the quotient is divisible by p. □

Problem 41. *Assume the integer $a \neq 1$ is not divisible by p and let $s \geq 1$. Prove*
$$p^s \mid a^{(p-1)p^{s-1}} - 1$$
by induction on s. Use the Little Fermat theorem and lemma 16.

Solution. For $s = 1$ the assertion is just Fermat's Little Theorem. Here is the induction step $s \to s + 1$:

We put $c - 1 := a^{(p-1)p^{s-1}} - 1$ which by induction assumption is divisible by p^s. We have to check divisibility by p^{s+1} for the expression

$$a^{(p-1)p^s} - 1 = c^p - 1 = \frac{c^p - 1}{c - 1} \cdot (c - 1)$$

By lemma 16 the first factor is divisible by p. By induction assumption the second factor is divisible by p^s Hence

$$p^{s+1} \mid a^{(p-1)p^s} - 1$$

as to be shown. □

Problem 42. *Assume that $a \neq 0, 1, -1$ and $m \geq 1, n \geq 0$. Confirm that*

(3.2.7) $$a^m - 1 \mid a^n - 1 \Leftrightarrow m \mid n$$

Solution. Assume $a^m - 1 \mid a^n - 1$ and $m, n \geq 1$. Formula (3.2.7) yields

$$a^m - 1 = \gcd(a^m - 1, a^n - 1) = a^{\gcd(m,n)} - 1 \mid a^m - 1 \quad \text{and hence}$$
$$a^m = a^{\gcd(m,n)} \quad \text{and} \quad m = \gcd(m, n) \quad \text{and finally } m \mid n$$

The converse is even easier to check. □

Proposition 27 (The little proposition plus). *I use the convention from mathematica that the greatest common divisor is always positive, except for $\gcd(0, 0) = 0$. Let $n, m \geq 0$ be integers. Let a be any integer. Let $2^s \| m$ and $2^t \| n$ be the highest powers of*

3.2. THE GEOMETRIC SERIES

two dividing m respectively n.

(3.2.8)
$$\gcd(a^m + 1, a^n + 1) = 2 \quad \text{if } s \neq t \text{ and } a \text{ is odd;}$$
(3.2.9)
$$\gcd(a^m + 1, a^n + 1) = 1 \quad \text{if } s \neq t \text{ and } a \text{ is even;}$$
(3.2.10)
$$\gcd(a^m + 1, a^n + 1) = a^{\gcd(m,n)} + 1$$
if $s = t$ and either $a \geq 0$ or m and n are both even;

(3.2.11)
$$\gcd(a^m + 1, a^n + 1) = -(a^{\gcd(m,n)} + 1)$$
if $s = t$ and $a < 0$ and either m or n is odd.

Proof. Take the case $s < t$. The case $a = 0$ may be checked directly. The case $a = 1$ is obvious since $\gcd(2,2) = 2$. The case $a = -1$ leads in formula (3.2.8) to m odd or even, but n always even. One get the valid formula $\gcd(2,2) = \gcd(0,2) = 2$. Assume $|a| \geq 2$. Let $g = \gcd(m,n)$ and put $b := a^g$, $m' = m/g$, $n' = n/g$, hence $\gcd(m',n') = 1$. Formula (3.2.4) yields at least

$$\gcd(a^m + 1, a^n + 1) = \gcd(b^{m'} + 1, b^{n'} + 1)$$
$$\text{divides } \gcd(b^{2m'} - 1, b^{2n'} - 1) = b^2 - 1$$

We use that $m' := m/g$ is odd but $n' := n/g$ is even. Since $b^2 - 1 \neq 0$, successively we get smaller remainders.

$$\gcd(b^{m'} + 1, b^2 - 1) = \gcd(b^{m'} + b^2, b^2 - 1)$$
$$= \gcd(b^{m'-2} + 1, b^2 - 1) = \ldots$$
$$= \gcd(b + 1, b^2 - 1) = b + 1$$
$$\gcd(b^{n'} + 1, b^2 - 1) = \gcd(b^{n'} + b^2, b^2 - 1)$$
$$= \gcd(b^{n'-2} + 1, b^2 - 1) = \ldots$$
$$= \gcd(b^2 + 1, b^2 - 1) = 2$$

Together we conclude

$$\gcd(a^m + 1, a^n + 1) = \gcd(b^{m'} + 1, b^{n'} + 1, b^2 - 1) \mid \gcd(b+1, 2)$$

Since
$$\gcd(b+1, 2) = \begin{cases} 2 & \text{if } b \text{ is odd} \\ 1 & \text{if } b \text{ is even} \end{cases}$$

using that b is odd if and only if a is odd yields the result (3.2.8).

Take the case $s = t$. We put $b := -(a^{2^s})$, and use that $m' := 2^{-s}m$ and $n' := 2^{-s}n$ are both odd. From the little proposition 26 is obtained $\gcd(b^{m'} - 1, b^{n'} - 1) = |b^{\gcd(m', n')} - 1|$. We get

$$\gcd(a^m + 1, a^n + 1) = \gcd(b^{m'} - 1, b^{n'} - 1) = |b^{\gcd(m', n')} - 1|$$
$$= |a^{\gcd(m,n)} + 1|$$

Assume $a \neq -1$. Obviously $a^{\gcd(m,n)} + 1 > 0$ iff either $a \geq 0$ or m and n are both even. One obtains formula (3.2.10).

From negating is obtained $a^{\gcd(m,n)} + 1 < 0$ iff $a \leq -2$ and either m or n is odd. One obtains formula (3.2.11).

In the case $a = -1$ and either m or n are odd, they are both odd since $s = t$ is assumed. We obtain from formula (3.2.11) the assertion $\gcd(0, 0) = 0$, the convention of mathematica. $\square$

Which case did we need to exclude? With the conventions $0^0 = 1$ and $\gcd(0, 0) = 0$, there are no cases to be excluded.

Chapter 4

Fermat and Euler

4.1 Little Fermat from Iterated Mappings

4.1.1 Moebius Inversion

Definition 13 (Moebius function). The *Moebius function* $\mu(n)$ is defined to be

- $\mu(n) = +1$ if n is square-free and has an even number of prime divisors;

- $\mu(n) = -1$ if n is square-free and has an odd number of prime divisors;

- $\mu(n) = 0$ if n is not square-free.

One puts $\mu(1) = 1$.

Proposition 28 (Moebius inversion). *Let $\mathcal{G}$ be any commutative group, written with addition as group operation. Let $n \in \mathbf{N} \mapsto A_n \in \mathcal{G}$ and $n \in \mathbf{N} \mapsto B_n \in \mathcal{G}$ be any functions. Let $n \geq 1$ be any integer.*

If for all divisors $d \mid n$ holds $A_d = \sum_{e \mid d} B_e$ *then holds*

for all divisors $d \mid n$: $B_d = \sum_{e \mid d} \mu\left(\dfrac{d}{e}\right) A_e = \sum_{e \mid d} \mu(e) A\left(\dfrac{d}{e}\right)$

Remark. Proposition 28 is as useful in an <u>multiplicatively</u> written version, too.

Lemma 17. *Let $\mathcal{G}$ be any commutative group or field. One needs a commutative and associative multiplication, the unit e and the inverse. One puts $g^0 = e$ for all group elements g. Let $n \in \mathbf{N} \mapsto A_n \in \mathcal{G}$ and $n \in \mathbf{N} \mapsto B_n \in \mathcal{G}$ be any functions. Let $n \geq 1$ be any integer.*

If for all divisors $d \mid n$ holds $\displaystyle A_d = \prod_{e \mid d} B_e$

then for all divisors $d \mid n$ holds $\displaystyle B_d = \prod_{e \mid d} A_e{}^{\mu(d/e)} = \prod_{e \mid d} A_{d/e}{}^{\mu(e)}$

Definition 14 (λ-function). We define the function $\lambda : \mathbf{N} \mapsto \mathbf{N}$ by setting

$$\lambda(n) = \begin{cases} p & \text{if } n = p^r \text{ is a prime power;} \\ 1 & \text{if } n = 1 \text{ or } n \text{ has at least two prime divisors.} \end{cases}$$

Lemma 18. *For all $n \geq 1$ holds*

(4.1.1) $$n = \prod_{d \mid n} \lambda(d)$$

Proof. Indeed that is easy to check for prime-powers. Define the sets
$$D_n = \{d \in \mathbf{N} : 1 \leq d \mid n \text{ and } d \text{ not prime-power}\}$$
Suppose there is a smallest counterexample. Let

$$n = \prod_i p_i^{r_i}$$

be its prime factorization. We already know that n is not a prime power. But simply calculate

$$\prod_{d \mid n} \lambda(d) = \prod_i \prod_{d = p_i^s \mid n} \lambda(d) \cdot \prod_{d \in D_n} \lambda(d) = \prod_i p_i^{r_i} = n$$

No smallest counterexample does exists, the assertion holds for all n □

Let me use the Kronecker delta

$$\delta_{ab} = \begin{cases} 1 & \text{if } a = b; \\ 0 & \text{if } a \neq b. \end{cases}$$

4.1. LITTLE FERMAT FROM ITERATED MAPPINGS

Lemma 19 (Remember Moebius). *The Moebius function satisfies for all $n \geq 1$*

(4.1.2) $$\sum_{d|n} \mu(d) = \delta_{n1}$$

(4.1.3) $$\sum_{d|n} \frac{n}{d} \mu(d) = \phi(n)$$

(4.1.4) $$\prod_{d|n} \left(\frac{n}{d}\right)^{\mu(d)} = \lambda(n)$$

Problem 43. *Obviously holds*

$$\sum_{d|n} \delta_{d1} = 1$$

Use Moebius inversion to get formula (4.1.2).

Problem 44. *By Gauss' proposition 47*

(5.3.1) $$\sum \{\phi(d) : d \text{ divides } m\} = m$$

Use Moebius inversion to get formula (4.1.3).

Problem 45. *We have proved above the formula (4.1.1). Use multiplicative Moebius inversion to get formula (4.1.4).*

4.1.2 Using an Iterated Mapping

Let $a \geq 1$ be an integer. Let $T : [0,1) \mapsto [0,1)$ be the transformation
$$T(x) = \lfloor ax \rfloor$$

Remark. One may close the half-open interval $[0,1)$ to a circle, use the topology for this circle and thus even get a continuous mapping T. That approach is more natural but does not touch the matter dealt with below.

The iterated mappings $T^{[n]}$ are defined recursively for all $n \geq 1$ by setting

$$T^{[1]} := T \text{ and } T^{[n+1]} := T^{[n]} \circ T \text{ for all } n \geq 1.$$

The reader should convince himself that the following facts are true.

Lemma 20. *Let $a \geq 1$. The iterated mapping are*

$$T^{[n]}(x) = \lfloor a^n x \rfloor \text{ for all } n \geq 1.$$

Let $a \geq 2$ and $n \geq 1$ be integers. The mapping $T^{[n]}$ has $a^n - 1$ fixed points. They occur at the points

$$x = \frac{l}{a^n - 1} \text{ for integers } 0 \leq l < a^n - 1.$$

For any $d \geq 1$ let b_d count the number of periodic <u>orbits</u> of $T^{[d[}$ with the <u>least</u> period d. For each divisor $d \mid n$ such a periodic orbit

$$x, T(x), T^{[2]}(x), \ldots,, T^{[d-1]}(x)$$

gives rise to d fixed points of $T^{[n]}$; Hence

Lemma 21. *Let $a \geq 2$ and $n \geq 1$ be integers. The number of fixed points of the mapping $T^{[n]}$ is*

$$\sum_{d \mid n} d \cdot b_d$$

Lemma 22 (Thanks to Moebius). *Let $a \geq 1$ and $n \geq 1$ be any integers.*

(4.1.5) $$\sum_{d \mid n} d\, b_d = a^n - 1$$

$$\sum_{d \mid n} \delta_{d1} + d\, b_d = a^n$$

(4.1.6) $$\sum_{d \mid n} \mu(d) a^{n/d} = n b_n + \delta_{n1}$$

Corollary 17 (Thanks to Moebius). *For any $a \geq 1$ and $n \geq 1$ the sum*

(4.1.7) $$\sum_{d \mid n} \mu(d)\, a^{\frac{n}{d}} \text{ is divisible by } n.$$

Too, this sum is positive or zero. It is zero if and only if $n \geq 2$ and no orbit of period n exists.

Corollary 18 (Fermat's Little Theorem). *For any integer a and any prime p holds $p \mid a^p - a$. If $p \nmid a$ then even $p \mid a^{p-1} - 1$ holds.*

4.1. LITTLE FERMAT FROM ITERATED MAPPINGS

Proof. Put $n = p$ and obtain for the sum (4.1.7)

$$\sum_{d|n} \mu(d)\, a^{\frac{n}{d}} = \mu(1)a^n - \mu(p)\, a^{\frac{n}{p}} = a \cdot [a^{p-1} - 1]$$

to be divisible by p. Under the assumption that $p \nmid a$, we conclude from Euclid's lemma that p divides the second factor $a^{p-1} - 1$. □

One may also get a few less obvious consequences.

Lemma 23. *For any integer a not divisible by p and any prime power p^s holds*

$$p^s \mid a^{(p-1)p^{s-1}} - 1$$

Proof. Put $n = p^s$ and obtain for the sum (4.1.7)

$$\sum_{d|n} \mu(d)\, a^{\frac{n}{d}} = \mu(1)a^n - \mu(p)\, a^{\frac{n}{p}} = a^{\frac{n}{p}}[a^{\frac{(p-1)n}{p}} - 1]$$

to be divisible by p^s. Under the assumption that $p \nmid a$, we conclude from Euclid's lemma that p^s divides the second factor $a^{(p-1)p^{s-1}} - 1$. □

Problem 46. *Let $p \neq q$ be two different prime and let a be an integer not divisible by neither p nor q. Prove that*

$$pq \mid a^{(p-1)(q-1)} - 1$$

Proof. Put $n = pq$ and obtain for the sum (4.1.7)

$$\sum_{d|n} \mu(d)\, a^{\frac{n}{d}} = \mu(1)a^n - \mu(p)\, a^{\frac{n}{p}} - \mu(q)\, a^{\frac{n}{q}} - \mu(pq)\, a^{\frac{n}{pq}}$$

$$= a^{pq} - a^q - a^p + a$$

which is divisible by pq. Since $\gcd(pq, a) = 1$ we get from the sum above, and from Little Fermat

$$pq \mid a^{pq-1} - a^{p-1} - a^{q-1} + 1$$
$$pq \mid (a^{p-1} - 1)(a^{q-1} - 1) = a^{p+q-2} - a^{p-1} - a^{q-1} + 1$$
$$pq \mid a^{pq-1} - a^{p+q-2} = a^{p+q-2} \cdot [a^{pq-p-q+1} - 1]$$
$$pq \mid a^{pq-p-q+1} - 1 = a^{(p-1)(q-1)} - 1$$

as to be shown. □

4.2 Euler's Totient Function

4.2.1 The Euler Group

Definition 15 (Unit group). For any commutative ring $\mathcal{R}$, its *unit group* consists of the elements having an inverse.

Problem 47. *Convince yourself that the unit group of any ring with unit is indeed a group.*

Definition 16 (Euler group). The *Euler group* consists of the remainder classes of a modulo m which are relatively prime to m. The modular multiplication is the group operation.

Lemma 24. *Fix some integer $m \geq 2$. The Euler group modulo m is the unit group of $\mathbf{Z}_m$.*

Proof. Let a be a unit in the ring $\mathbf{Z}_m$. There exists an inverse $b = a^{-1}$ for which holds $ab \equiv 1 \pmod{m}$. Hence

$$ab + km = 1$$

which implies that $\gcd(a, m) = 1$ and a is an element of the Euler group.

Conversely, let a be an element of the Euler group. By definition holds $\gcd(a, m) = 1$. By the extended Euclidean algorithm there exist integers b and k such that

$$ab + km = 1$$

Hence $ab \equiv 1 \pmod{m}$ and $b = a^{-1}$ in the ring $\mathbf{Z}_m$. Thus a is a unit in the ring $\mathbf{Z}_m$. $\square$

We denote the Euler group by G_m^* or $\mathcal{U}(\mathbf{Z}_m)$. The order of the Euler group

$$|\mathcal{U}(\mathbf{Z}_m)| = \phi(m)$$

is the Euler totient function.

4.2.2 The Euler Totient Function

Definition 17. For any $m > 1$, the *Euler totient function* $\phi(m)$ counts the number of integers among $1, 2, \ldots, m - 1$ which are relatively prime to m. Equivalently, one can count residue classes. One puts $\phi(1) = 1$.

4.2. EULER'S TOTIENT FUNCTION

Moreover, consider for the Chinese remainder problem

(ch) $\qquad x \equiv u \pmod{a}$ and $x \equiv v \pmod{b}$

the case that $\gcd(u, a) = \gcd(v, b) = 1$, and $\gcd(a, b) = 1$ as was assumed before. Under these assumptions, $\gcd(f(u,v), ab) = 1$ holds for the solution $f(u, v)$ of problem (ch). The converse is true, too: $\gcd(f(u,v), ab) = 1$ implies $\gcd(u, a) = \gcd(v, b) = 1$. Hence the restriction of the solution function $f : (u, v) \mapsto x$ to the residue classes $(u, v) \in \mathcal{U}(\mathbf{Z}_a) \times \mathcal{U}(\mathbf{Z}_b)$ leads to an interesting result about the structure of the group $\mathcal{U}(\mathbf{Z}_{ab})$:

Corollary 19. *For any relatively prime a and b we have a group isomorphism,*

$$\mathcal{U}(\mathbf{Z}_a) \times \mathcal{U}(\mathbf{Z}_b) \simeq \mathcal{U}(\mathbf{Z}_{ab})$$

obtained via a restriction the bijection from the Chinese Remainder Theorem 8. Hence the Euler totient function is multiplicative. In other words,

(4.2.1) $\qquad \phi(ab) = \phi(a)\phi(b)$

holds for any relatively prime integers a and b.

Definition 18 (Multiplicative function). A number theoretic function $n \in \mathbf{N} \mapsto f(n) \in \mathbf{Z}$ is called *multiplicative* if and only if

(4.2.2) $\qquad f(ab) = f(a) \cdot f(b)$

holds for any relatively prime a and b.

Problem 48. *Convince yourself that the Moebius function μ given by definition 13 is multiplicative.*

Problem 49. *Convince yourself that the Lambda function λ given by definition 14 is <u>not</u> multiplicative.*

Problem 50. *Show that for any prime power p^r the Euler totient function takes the value $\phi(p^r) = p^{r-1}(p-1)$.*

Proposition 29 (Formula for the totient function). *The Euler totient function takes the values:*

$$\phi(1) = \phi(2) = 1, \; \phi(2^r) = 2^{r-1} \quad \text{for } r \geq 1,$$
$$\phi(p^r) = p^{r-1}(p-1) \quad \text{for any odd prime } p \text{ and } r \geq 1$$
$$\phi(p_1^{r_1} \cdot p_2^{r_2} \cdots p_s^{r_s}) = \phi(p_1^{r_1}) \cdot \phi(p_2^{r_2}) \cdots \phi(p_s^{r_s})$$
$$= p_1^{r_1-1}(p_1-1) \cdot p_2^{r_2-1}(p_2-1) \cdots p_s^{r_s-1}(p_s-1)$$

where $p_1, p_2, \ldots p_s$ are any different primes.

Lemma 25. *The Euler totient function satisfies $\phi(n) \geq \sqrt{n}$ for all $n \neq 2, 6$. Equality holds only for $n = 1$ and $n = 4$.*

Proof. One gathers enough cases to get the general result.

- In the case that $n = p^r$ where $r \geq 1$ and $p \geq 5$ is an odd prime
$$\frac{\phi(n)^2}{n} = (p-1)^2 p^{r-2} \geq \frac{(p-1)^2}{2p} \geq 1.6$$
since $(p-1)^2 - 3 \cdot 2p = p^2 - 5 \cdot 2p + 1 \geq 0$ for $p \geq 5$.

- In the case that $n = 2^s p^r$ where $s \geq 1, r \geq 1$ and $p \geq 5$ is an odd prime
$$\frac{\phi(n)^2}{n} = 2^{s-2}(p-1)^2 p^{r-2} \geq \frac{(p-1)^2}{2p} \geq 1.6$$
since $(p-1)^2 - 3 \cdot 2p = p^2 - 5 \cdot 2p + 1 \geq 0$ for $p \geq 5$.

- In the case that $n = 2^r$ with $r \geq 2$
$$\frac{\phi(n)^2}{n} = 2^{2(r-1)-r} \geq 1$$

- In the case that $n = 3^r$ with $r \geq 1$
$$\frac{\phi(n)^2}{n} = 2^2 \cdot 3^{2(r-1)-r} \geq \frac{4}{3} > 1$$

- In the case that $n = 2^s \cdot 3$ with $s \geq 2$
$$\frac{\phi(n)^2}{n} = 2^{2(s-1)+2-s} \cdot 3^{-1} \geq \frac{4}{3} > 1$$

- In the case that $n = 2^s 3^r$ with $s \geq 1, r \geq 2$
$$\frac{\phi(n)^2}{n} = 2^{2(s-1)+2-s} \cdot 3^{2(r-1)-r} \geq 2 > 1$$

The remaining cases may be covered by multiplicativity (4.2.1). □

Problem 51 (Schinzel 1956). *Show that the Euler totient function never assumes any of the values $2 \cdot 7^k$ for $k \geq 1$.*

(i) *Show that all numbers $1 + 2 \cdot 7^k$ for $k \geq 1$ are divisible by 3 and hence composite.*

4.2. EULER'S TOTIENT FUNCTION

(ii) Conclude that $\phi(p) \neq 2 \cdot 7^k$ for all primes p and $k \geq 1$.

(iii) Convince yourself that $\phi(2p^r) = \phi(p^r) \neq 2 \cdot 7^k$ for all odd prime-powers p with $r \geq 1$ and $k \geq 1$.

(iv) Show that 4 divides $\phi(m)$ if $m \geq 5$ and $m \neq p^r, m \neq 2p^r$ is neither an odd prime-power nor twice an odd prime power.

(v) Conclude that the Euler totient function never assumes any of the values $2 \cdot 7^k$ for $k \geq 1$.

Answer. (i) Let $N := 1 + 2 \cdot 7^k$ and $k \geq 1$. Calculate modulo 3 to get $N \equiv 1 - 1^k = 0$ and hence N is divisible by three and composite.

(ii) Obviously $\phi(2), \phi(3) \neq 2 \cdot 7^k$ for any $k \geq 1$. Given a prime $p \geq 5$ the value $p - 1 \neq 2 \cdot 7^k$ for any $k \geq 1$ since $1 + 2 \cdot 7^k$ is composite.

(iii) Suppose towards a contradiction that $p^{r-1}(p-1) = 2 \cdot 7^k$ for any prime-power of p with $r \geq 1$ and $k \geq 1$. It is impossible to have $p = 2, 3, 5$ since no factor 7 would occur in $\phi(p^r)$. What about $p = 7^r$ and $r \geq 1$. In that case we get

$$\phi(7^r) = 6 \cdot 7^{r-1} \neq 2 \cdot 7^k$$

because of the uniqueness of the prime factorization. Consider the case $p \geq 11$. Since no factor p does occur, we get $r = 1$. That case was already ruled out in part (i).

(iv) Can be left to the reader. [1]

(v) The Euler totient function $\phi(m)$ never assumes any of the values $2 \cdot 7^k$ for $k \geq 1$. Neither can m be equal to 1, nor to a prime power nor to twice a prime power. Nor can m be another composite number since for these the value $\phi(m)$ is divisible by 4.

Recursive method to find solutions of $\phi(m) = c$ for given positive integer c

(i) $c = 1$: Two solutions $m = 1$ or $m = 2$. No further solutions exist. The recursive program stops.

(ii) $c \geq 3$ **odd**: No solution exists. The recursive program stops.

---
[1] See Problem 72 below.

(iii) **"flower"** Check whether $c+1$ is odd prime. If yes, two more solutions $m = c + 1$ or $m = 2(c+1)$ exist.

(iv) **"leaf"** Prime factor c and let p be the largest prime occurring. Define the exponent r by $p^{r-1} \| c$. Check whether $(p-1)p^{r-1} = c$. If yes, two more solutions exist: $m = p^r$ or $m = 2p^r$.

(v) **"exhausted"** $c \equiv 2 \pmod 4$: If $c = 2$, one gets the solution $m = 4$, additionally to $3, 6$ obtained in item (iii). Otherwise, no further solutions exist besides those already obtained in items (iii) and (iv). The recursive program has to return.

(vi) **"branching"** $c \equiv 0 \pmod 4$: Find all products $c = ab$ with $2 \le a \le b$, and a and b both even. In each case, solve $\phi(l) = a$ and $\phi(k) = b$, applying the procedure recursively. For each pair with $\gcd(l, k) = 1$, one gets a solution $m = lk$. In the end, one needs to eliminate solutions occurring more than once.

Problem 52. *Find six values of m for which $\phi(m) = 12$.*

Answer. $\phi(13) = \phi(26) = 12$ since $12 + 1$ is prime. To find more values, put $m = kl$ with $\gcd(k, l) = 1$ and look for solutions of $\phi(k) = 2, \phi(l) = 6$. We know the solutions $k = 3, 4, 6$ and $l = 7, 14, 9, 18$ from above. One needs to select the pairs were a and b are relatively prime and delete solutions occurring twice. Thus four extra solutions remain. :

$$3 \cdot 7 = 21,\ 3 \cdot 14 = 42,\ 4 \cdot 7 = 28,\ 4 \cdot 9 = 36$$

Altogether we have six solutions

$$3 \cdot 7 = 21,\ 3 \cdot 14 = 42,\ 4 \cdot 7 = 28,\ 4 \cdot 9 = 36,\ 13,\ 26$$

Problem 53. *Find seven values of m for which $\phi(m) = 12 \cdot 13$.*

Answer.

(iii) **"flower"** Check whether $12 \cdot 13 + 1 = 157$ is odd prime. Yes, hence there are two solutions $m_1 = 157$ and $m_2 = 314$.

(iv) **"leaf"** Prime factor $c = 2^2 \cdot 3 \cdot 13$. We see that 13 be the largest prime occurring, with exponent $r = 1$. Indeed $12 \cdot 13 = (p-1)p^{r-1} = c$. Hence two more solutions exist: $m_3 = 13^2 = 169$ and $m_4 = 2p^r = 338$.

4.2. EULER'S TOTIENT FUNCTION

(vi) "branching" c is divisible by 4 Find all products $c = ab$ with $2 \leq a \leq b$, and a and b both even. One gets two solutions $12 \cdot 13 = 2 \cdot 78 = 6 \cdot 26$. One checks that $\phi(m) \neq 26$, and the second decomposition does not lead to further solutions.

But the first product leads to further solutions. Indeed, $\phi(l) = 2$ holds for $l = 3, 4, 6$ and $\phi(k) = 78$ holds for $k = 79, 2 \cdot 79$. The relatively prime pairs $\gcd(l, k) = 1$ are $m_5 = 3 \cdot 79 = 237, m_6 = 4 \cdot 79 = 316, m_7 = 6 \cdot 79 = 474$.

All solutions There are seven solutions
$m_i = 157, 314, 169, 338, 237, 316, 474$.

Remark. From the last problem we get an example of two numbers $m, m' > 1$ such that $\phi(m) = \phi(m')$, and a primitive root exists modulo m, but no primitive root exists modulo m'. One can take for example $m = 157$ and $m' = 237$.

DrRacket program for the recursive method to find all solutions of $\phi(m) = c$ for given positive integer c

```
(require math/base)
(require math/number-theory)

(define (primepower-solutions c)
  (cond [(equal? c 1)'(1 2)]
        [(odd? c) null]
        [(equal? c 2)'(3 4 6)]
;       [(equal? c 4)'(5 8 10 12)]
;       [(equal? c 6)'(9 18 7 14)]
;       [(equal? c 8)' (15 16 20 24 30)]
        [(power-of-two? c)
           (if (prime? (add1 c))
               (list (* c 2) (add1 c) (* 2 (add1 c)))
               (list (* c 2)))]
        [else
          (let*[(found1 (if (prime? (add1 c))
                 (list (add1 c) (* 2 (add1 c)))null))
                (p (last (prime-divisors c)))
                (rm1(last (prime-exponents c)))
                (ctry(*(sub1 p)(expt p rm1)))
                (pr(expt p(add1 rm1)))
                (news (list pr (* 2 pr)))
                (result (if (equal? c ctry)
```

```
                    (append news found1)
                     found1))]
              result)])))

(define(decomposed c)
   (if (divides? 4 c)
      (let*[(divs (divisors (/ c 4)))
            (counted (floor(/(length divs)2)))
            (bshalf (list-tail divs counted))]
       (map (lambda(b)(* 2 b)) bshalf))null))

(define filter-cart-multiply
   (lambda(a b)
     (let* [(good-pairs
              (filter
                (lambda(item)
                  (equal? 1 (gcd(car item)(cadr item))))
                (cartesian-product a b)))]
         (map (lambda(item)
                  (* (car item)(cadr item)))
               good-pairs))))

(define (list-union lsts)
   (cond [(null? lsts) null]
    [(null? (car lsts)) (list-union (cdr lsts))]
    [else (append (car lsts)(list-union (cdr lsts)))]))

(define (phi-inverse c)
   (if (divides? 4 c)
      (let*[(dc (decomposed c))
            (exes (map (lambda(twob)
                     (filter-cart-multiply
                        (phi-inverse (/ c twob))
                        (phi-inverse twob))) dc))
            (together (append (list-union exes)
                                (primepower-solutions c)))]
         (sort (remove-duplicates together)<))
      (primepower-solutions c)))

(let* [(c 1024)
       (phim (phi-inverse c))]
 (displayln phim)
```

4.2. EULER'S TOTIENT FUNCTION

```
(displayln(map totient phim)))
```

Remark. The lines

```
;       [(equal? c 4)'(5 8 10 12)]
;       [(equal? c 6)'(9 18 7 14)]
;       [(equal? c 8)' (15 16 20 24 30)]
```

are not needed but may accelerate the computations. The last four lines give an example, and its check.

Problem 54. *Find c for which $\phi(m) = c$ has exactly two solutions.*

Proof. $\phi(m) = 52$ has only two solutions $m = 53$ and $m = 106$. □

Problem 55. *Find some solution of $\phi(m) = 4 \cdot 3^r$ with arbitrary integer $r \geq 1$. Convince yourself that there exist at least three solutions.*

Proof. $\phi(4 \cdot 3^{r+1}) = 4 \cdot 3^r$ and $\phi(7 \cdot 3^r) = \phi(14 \cdot 3^r) = 4 \cdot 3^r$ □

Remark. The equation $\phi(m) = 4 \cdot 3^r$ has for $r = 0, 1, \ldots 34$ the following number of solutions

(4, 6, 8, 9, 8, 12, 18, 15, 13, 16, 17, 17, 13, 15, 17, 17, 16, 20, 21, 19, 19, 21, 21, 19, 17, 19, 19, 17, 17, 17, 20, 21, 21, 21, 21)

Problem 56. *Find all solution of $\phi(m) = 4 \cdot 13^r$ with arbitrary integer $r \geq 1$. Convince yourself that there exist no solutions if $1 + 4 \cdot 13^r$ is composite. How many solution are there if $1 + 4 \cdot 13^r$ is prime.*

Solution. If $1 + 4 \cdot 13^r$ is composite, the equation $\phi(n) = 4 \cdot 13^r$ has no prime-power solutions. Further "branching" solutions can only exist if the equation $\phi(a) = 2 \cdot 13^s$ is solvable for some $0 < s \leq r$. This is only the case if $1 + 2 \cdot 13^s$ is prime. But this is impossible since 3 is a divisor of $1 + 2 \cdot 13^s$. Hence the equation $\phi(n) = 4 \cdot 13^r$ has neither prime-power solutions nor "branching" solution.

If $1 + 4 \cdot 13^r$ is prime, the equation $\phi(n) = 4 \cdot 13^r$ has the two solutions $1 + 4 \cdot 13^r$ and $2(1 + 4 \cdot 13^r)$. □

Remark. The number of solutions of $\phi(n) = 4 \cdot 3^i$ for $i = 0, 1 \ldots 49$ is

$$4, 6, 8, 9, 8, 12, 18, 15, 13, 16, 17, 17, 13, 15, 17, 17, 16, 20, 21, 19, 19,$$
$$21, 21, 19, 17, 19, 19, 17, 17, 17, 20, 21, 21, 21, 21, 21, 21, 19, 19, 23,$$
$$19, 19, 19, 19, 19, 19, 21, 21, 19, 19$$

The number of solutions of $\phi(n) = 4 \cdot 5^i$ for $i = 0, 1, \ldots, 49$ is

$$4, 5, 4, 5, 4, 2, 4, 2, 2, 2, 2, 2, 2, 5, 4, 2, 4, 2, 4, 2, 2, 2, 2, 2, 2, 2,$$
$$2, 2, 2, 2, 2, 2, 2, 2, 2, 2, 2, 2, 2, 2, 2, 2, 2, 2, 5, 4, 2, 4, 2,$$

The number of solutions of $\phi(n) = 4 \cdot 7^i$ for $i = 0, 1, \ldots, 49$ is

$$4, 2, 2, 2, 0, 0, 2, 2, 0, 0, 2, 2, 0, 0, 0, 0, 0, 2, 0, 0, 0, 0, 0, 0, 0,$$
$$0, 0, 0, 0, 0, 0, 0, 0, 0, 0, 0, 0, 2, 0, 0, 0, 0, 0, 0, 0, 0, 0, 0, 0,$$

The number of solutions of $\phi(n) = 4 \cdot 11^i$ for $i = 0, 1, \ldots, 49$ is

$$4, 3, 0, 3, 2, 0, 0, 0, 0, 3, 2, 0, 2, 0, 0, 0, 0, 0, 0, 0, 0, 0, 0, 0, 0, 0,$$
$$0, 0, 0, 0, 0, 0, 0, 0, 0, 0, 0, 0, 0, 0, 0, 3, 2, 0, 2, 0, 0, 0,$$

The number of solutions of $\phi(n) = 4 \cdot 13^i$ for $i = 0, 1, \ldots, 49$ is

$$4, 2, 2, 0, 0, 0, 0, 0, 0, 0, 0, 0, 0, 0, 0, 0, 0, 0, 2, 0, 0, 0, 0, 0, 0, 0,$$
$$0, 0, 2, 0, 0, 0, 0, 0, 0, 0, 2, 0, 0, 0, 0, 0, 0, 0, 0, 0, 0, 0, 0,$$

The number of solutions of $\phi(n) = 4 \cdot 17^i$ for $i = 0, 1, \ldots, 49$ is

$$4, 0, 0, 0, 0, 0, 2, 0, 0, 0, 0, 0, 0, 0, 0, 0, 0, 0, 0, 0, 0, 0, 0, 0, 0, 0,$$
$$0, 0, 0, 0, 0, 0, 0, 0, 0, 0, 0, 0, 0, 0, 0, 0, 0, 0, 0, 0, 3, 0, 0,$$

The number of solutions of $\phi(n) = 4 \cdot 19^i$ for $i = 0, 1, \ldots, 49$ is

$$4, 0, 0, 2, 0, 2, 0, 0, 0, 0, 0, 0, 0, 0, 0, 0, 0, 0, 0, 0, 0, 0, 0, 0, 0, 0,$$
$$0, 0, 0, 0, 0, 0, 0, 0, 0, 0, 0, 0, 0, 0, 0, 0, 0, 0, 0, 0, 0, 0, 0,$$

Problem 57. *Find some conjectures based on the last remark 4.2.2.*

Remark. A longstanding open conjecture of Carmichael says that the equation $\phi(m) = c$ has either no or at least two solutions. (see article [9] by Wagon, *The Mathematical Intelligencer*, 1986.) The conjecture has been checked to be true for $\phi(m) < 10^{10000}$ as of ca. 1985.

Remark. Try the case $c = 4 \cdot 3^r$ with $r \geq 1$. Of course $\phi(4 \cdot 3^{r+1}) = 4 \cdot 3^r$. Hence $m_1 = 4 \cdot 3^{r+1}$ is a solution occurring for all $r \geq 1$. If $s = 1 + 4 \cdot 3^r$ is prime, we get two more solutions $s, 2s$ as a "flower". Too, if $q = 1 + 2 \cdot 3^r$ is prime, a "branching" occurs, and we get three more solutions $3q, 4q, 6q$. There are many further recursive branchings and hence more solutions.

4.2.3 Euler's Theorem

Definition 19. Let $m > 1$ and $\gcd(a, m) = 1$. The *order of an integer a modulo m* is the smallest positive integer ω such that

$$a^\omega \equiv 1 \pmod{m}$$

We denote the order of an integer a modulo m by $\operatorname{ord}_m(x)$.

Note that $a, a^2, \ldots a^\omega \equiv 1$ are the elements of a cyclic subgroup of $\mathcal{U}(\mathbf{Z}_m)$. By Lagrange's Theorem, the order of a subgroup is a divisor of the order of the group. Hence we conclude

Proposition 30. *Let $m > 1$ and $\gcd(a, m) = 1$. The order of an integer a modulo m is a divisor of the Euler totient function $\phi(m)$.*

Theorem 9 (Euler's Theorem).

$$a^{\phi(m)} \equiv 1 \pmod{m}$$

for any $m > 1$ provided that $\gcd(a, m) = 1$.

Independent proof of Euler's Theorem. Let $b_1, b_2, \ldots, b_{\phi(m)}$ be representatives for all the residue classes relatively prime to m and put

$$A := b_1 \cdot b_2 \cdots b_{\phi(m)}$$

For an integer a with $\gcd(a, m) = 1$, the numbers $ab_1, ab_2 \ldots ab_{\phi(m)}$ are again representatives for all the residue classes relatively prime to m. In other words, as elements of $\mathbf{Z}_m$, they are a permutation of $b_1, b_2, \ldots, b_{\phi(m)}$. We conclude

$$a^{\phi(m)} A = ab_1 \cdot ab_2 \cdots ab_{\phi(m)} \equiv b_1 \cdot b_2 \cdots b_{\phi(m)} = A$$
$$(a^{\phi(m)} - 1) A \equiv 0$$

Since A is relatively prime to m, Euclid's Lemma implies m divides $a^{\phi(m)} - 1$ and hence $a^{\phi(m)} \equiv 1 \pmod{m}$ as to be shown. □

A second proof of Euler's Theorem. For the special case that $m = p^s$ is a prime power we may use lemma 23. Since the integer a is assumed to be relatively prime to p^s, it is not divisible by p. Hence

$$p^s \mid a^{(p-1)p^{s-1}} - 1 = a^{\phi(p^s)} - 1$$

Let any relatively prime integers a and m with $\gcd(a, m) = 1$ be given. By the little proposition 26 we know that $d \mid \phi(m)$ implies $a^d - 1 \mid a^{\phi(m)} - 1$. Let

$$m = \prod_i p_i^{r_i}$$

be the prime factorization. Since for any index i we know that $\phi(p_i^{s_i}) \mid \phi(m)$, one gets for all i

$$p_i^{s_i} \mid a^{\phi(p_i^{s_i})} - 1 \mid a^{\phi(m)} - 1$$

Since these prime powers are relatively prime, we may conclude

$$m = \prod_i p_i^{s_i} \mid a^{\phi(m)} - 1$$

as to be shown. □

Proposition 31. *Given are any integer $a \neq 1, -1$, integer $m > 1$ and a prime p. If $m \mid a^p - 1$ then either $p \mid \phi(m)$ or $m \mid a - 1$.*

Proof. By Euler's theorem $m \mid a^{\phi(m)} - 1$. Hence the little proposition 26 yields

$$m \mid \gcd(a^{\phi(m)} - 1, a^p - 1) = a^{\gcd(\phi(m), p)} - 1$$

In the case that $p \nmid \phi(m)$ we get $\gcd(\phi(m), p) = 1$ since p is prime. Hence $m \mid a - 1$. □

Proposition 32. *Given are any integer $a \neq 1, -1$, and two primes p and q. If $q \mid a^p - 1$ then either $q = 1 + kp$ or $m \mid a - 1$. If p and q are odd, we conclude even $q = 1 + 2kp$ or $m \mid a - 1$.*

Definition 20 (Mersenne prime). Prime p is called a *Mersenne prime* if and only if $2^p - 1$ is a prime.

Definition 21 (Sophie-German prime). The prime p is called a *Sophie-German prime* if and only if both p and $2p+1$ are primes.

4.2. EULER'S TOTIENT FUNCTION

Problem 58. *Find all solutions of the equation*

(4.2.3) $$\phi(n+2) = \phi(n) + 2$$

Let p denote any prime. Show that

(i) $n = p$ *is a solution of equation* (4.2.3) *if and only if p and $p+2$ are twin primes.*

(ii) $n = 2p$ *is a solution of equation* (4.2.3) *if and only if p is an odd Mersenne prime.*

(iii) $n = 4p$ *is a solution of equation* (4.2.3) *if and only if p is an odd Sophie-German prime.*

Find a solution of equation (4.2.3) *with $n \leq 30$ not among the above ones.*

Conjecture 3. *Equation* (4.2.3) *has only the one exceptional solution $n = 18$ not covered by items* (i),(ii) *or* (iii). *I have checked the conjecture for all $n \leq 10^6$.*

My unsuccessful attempt to prove conjecture 3 has only produced the following partial result.

Proposition 33. *Any solution of equation* (4.2.3) *for which both n and $n+2$ are of the form $2^a p^b$ where p is prime is either one covered by items* (i),(ii) *or* (iii) *or the exceptional solution $n = 18$.*

Proof. Take the case that n is odd. Hence the additional assumption tells that $n = p^j$ and $n+2 = q^k$ are odd primes or prime powers. With this Ansatz equation (4.2.3) becomes

$$(q-1)q^{k-1} = 2 + (p-1)p^{j-1}$$

Hence $q^{k-1} = p^{j-1}$. Now $p \neq q$ are different odd primes since $\gcd(n+2, n) = 2$. The uniqueness of prime decomposition implies $k = j = 1$. Hence $q = p+2$ and we have arrived at the case of item (i).

The case that n is even is solved in lemma 26 and 27 below. □

Lemma 26. *Any even exceptional solution of equation* (4.2.3) *may be written as*

(4.2.4) $$\phi(n + \sigma 2) = \phi(n) + \sigma 2 \quad \text{with } \sigma = \pm 1 \text{ and } 2 \| \phi(n).$$

Hence

$$n = 2p^j \quad \text{where } p \equiv 3 \mod 4 \text{ is an odd prime and } j \geq 2.$$

Moreover

$$m := \frac{n + \sigma 2}{4}$$

is an integer. Finally $p + 2 \leq q_i$ holds for all odd prime divisors of $q_i \mid m$.

Proof. Define Q by setting

$$\phi(n + \sigma 2) = \frac{n + \sigma 2}{2}\left(1 - \frac{1}{Q}\right)$$

The equation (4.2.4) to be solved may be written as

$$1 - \frac{1}{Q} = \frac{(p-1)p^{j-1} + \sigma 2}{p^j + \sigma} \quad \text{or equivalently}$$

(4.2.5) $$Q = p + \sigma \frac{p+1}{p^{j-1} - \sigma}$$

By the formula for the totient function from proposition 29 holds

$$\frac{\phi(n + \sigma 2)}{n + \sigma 2} = \frac{1}{2}\left(1 - \frac{1}{Q}\right) = \frac{1}{2}\prod\left(1 - \frac{1}{q_i}\right)$$

where $q_i \mid m$ denote the different odd prime divisors. Let $q_1 \mid m$ be the smallest <u>odd</u> prime divisor. Hence

$$1 - \frac{1}{Q} \leq 1 - \frac{1}{q_1} \quad \text{and} \quad Q \leq q_1$$

Moreover $Q = q_1$ if and only if the odd part of m is a prime power. With the value from equation (4.2.5) for Q and $p \geq 3$, $j \geq 2$ one gets

(4.2.6) $$p - 1 \leq p + \sigma\frac{p+1}{p^{j-1} - \sigma} = Q \leq q_1$$

Now $p \neq q_i$ are different odd primes since $\gcd(m, n) \leq 2$. Hence $p + 2 \leq q_i$ holds for all odd prime divisors $q_i \mid m$. □

Remark. The special case $p = 3, q_1 = 2$ leads to $\sigma = -1, j = 2, p = 3$ $n = 18, m = 8$ and $8 = \phi(16), 6 = \phi(18)$. Not a solution of equation (4.2.3). I could not even prove that $j \geq 2$ implies $m \equiv 4 \mod 8$.

4.2. EULER'S TOTIENT FUNCTION

Lemma 27. *There is only one even exceptional solution of equation (4.2.3) for which the odd parts of both n and $n+2$ are prime powers: this is $n = 18$.*

Proof. Define Q as in the previous lemma 27. Under the additional assumption that the odd parts of both n and $n+2$ are prime powers one gets $Q = q_1$ which is an odd prime. With the value for Q from equation (4.2.6) and $p \geq 3$, $j \geq 2$ one gets

$$p + \sigma \frac{p+1}{p^{j-1} - \sigma} = Q = q_1$$

Too, we know from the previous lemma that $p + 2 \leq Q$. Hence

$$p + 2 \leq p + \sigma \frac{p+1}{p^{j-1} - \sigma} = Q \text{ is an odd prime.}$$

We conclude that $\sigma = +1$ and $j = 2$. From $2 \leq \frac{p+1}{p-1}$ one finally gets $p = 3$. Hence $n = 2p^j = 18$ as claimed. □

Definition 22 (Carmicheal function). The *Carmicheal function*, denoted by $\lambda(m)$ was defined in 1912 with the values:

$$\lambda(1) = \lambda(2) = 1, \ \lambda(4) = \lambda(8) = 2,$$
$$\lambda(2^r) = \phi(2^r)/2 = 2^{r-2} \quad \text{for } r \geq 3,$$
$$\lambda(p^r) = \phi(p^r) = p^{r-1}(p-1) \quad \text{for any odd prime } p \text{ and } r \geq 1$$
$$\lambda(p_1^{r_1} \cdot p_2^{r_2} \cdots p_s^{r_s}) = \text{lcm}\,[\lambda(p_1^{r_1}), \lambda(p_2^{r_2}), \ldots, \lambda(p_s^{r_s})]$$

where $p_1, p_2, \ldots p_s$ are any different primes.

Problem 59. *Convince yourself that the Carmichael function λ given by definition 22 is not multiplicative.*

Theorem 10 (Carmichael function). *For any $m > 1$, the Carmichael function is the lowest exponent such that*

$$a^{\lambda(m)} \equiv 1 \pmod{m}$$

for all integers a which are $\gcd(a, m) = 1$ relatively prime to m.

For all divisors $d \mid \lambda(m)$ of the Carmichael function, there exists an integer b such that d is equal to the order $\text{ord}_m(b)$ of b modulo m.

End of the proof of Theorem 10. Given is the integer $m > 1$. As shown in Lemma 59, for any integer a, the order modulo m is a divisor of the Carmichael function $\lambda(m)$. As shown in Proposition 53, there exists an integer g for which the order modulo m is equal to the Carmichael function $\lambda(m)$. Hence the Carmichael function is the lowest exponent being for all integers a a multiple of their respective orders modulo m. □

Problem 60. *Calculate $\phi(5040)$ and $\lambda(5040)$. Find a polynomial equation of degree* 12 *with more than* 1000 *solutions. How many roots has the polynomial $x^{12} - 1 \in \mathbf{Z}_{5040}[x]$.*

Answer. $5040 = 2 \cdot 3 \cdot 4 \cdot 5 \cdot 6 \cdot 7 = 2^4 \cdot 3^2 \cdot 5 \cdot 7$. We calculate $\phi(5040) = 2^3 \cdot 2 \cdot 3 \cdot 4 \cdot 6 = 128 \cdot 9 = 1152$ and $\lambda(5040) = \mathrm{lcm}(2^2, 6, 4, 6) = 12$.

The polynomial $x^{12} - 1 \in \mathbf{Z}_{5040}[x]$ has 1152 different roots relatively prime to 5040. On the other hand if $\gcd(x, 5040) > 1$ the $p \mid x$ for one of the primes $p = 2, 3, 5, 7$. Hence $x^{12} \equiv 0$ (mod p) contradicting $x^{12} \equiv 1$ (mod 5040). The polynomial $x^{12} - 1 \in \mathbf{Z}_{5040}[x]$ has no roots that are not relatively prime to 5040.

Chapter 5

On Giants Shoulders

The integers are denoted by **Z**. For an *indeterminant* or *variable* x, the expressions $1, x, x^2, x^3, \ldots$ are called monomials. An integer combination of monomials $a_n x^n + a_{n-1} x^{n-1} + \cdots + a_1 x + a_0$ with integer coefficients $a_n, a_{n-1}, \ldots, a_0 \in \mathbf{Z}$ is called an *integer polynomial* or *polynomial over the integers*. The set of all integer polynomials is denoted by $\mathbf{Z}[x]$. One has to put the *indeterminant into square brackets*.

The set of integers **Z** and the set $\mathbf{Z}[x]$ are the first examples for a *ring*. The reader should recall the definition of a ring.

Definition 23 (ring). A *ring* is a collection of elements $a, b, \ldots$ which contains the neutral element 0 and allows two binary operations $+$ and $\cdot$ such that for all a, b in the ring, the sum $a + b$ and the product $a \cdot b = ab$ are elements of the ring, too. Moreover for all a, b, c in the ring hold the following arithmetic rules:

neutral element $0 + a = a$

negative element for all a exists the negative element $-a$ such that $a + (-a) = 0$.

commutativity $a + b = b + a$

associativity $(a + b) + c = a + (b + c)$

left distributivity $(a + b)c = ac + bc$

right distributivity $c(a + b) = ca + cb$

For any ring R, we can define the polynomial ring $R[x]$ by taking elements [1] One uses $a_n, a_{n-1}, \ldots, a_0 \in R$ as coefficients in the respective polynomials. Again, provided that R is a ring, the set of polynomials $R[x]$ is a ring, too. It is called the *polynomial ring over the ring R*.

A polynomial with the leading monomial $a_n x^n$ and $a_n \neq 0$ is defined to have the *degree* n. A polynomial with the leading monomial x^n is called *monic*.

The polynomials of *positive degree* are those with $n \geq 1$ and hence those in which the indeterminant x really appears.

The polynomials with degree 0 are the elements $a_0 \neq 0$ of the ring R. The zero polynomial has the degree undefined or $-\infty$. [2]

Remark. Anyway holds $\deg 0 \neq 0$. Some authors even write $\deg P \geq 0$ to indicate that the polynomial $P \neq 0$.

Definition 24 (field). A *field* is a collection of elements $a, b, c \ldots$ which contains the neutral element 0, the unit element 1, and allows two binary operations $+$ and $\cdot$ such that for all a, b in the field, the sum $a + b$ and the product $a \cdot b = ab$ are elements of the field, too. Moreover for all a, b, c hold the following arithmetic rules:

neutral element $0 + a = a$

negative element for all a exists the negative element $-a$ such that $a + (-a) = 0$.

commutativity $a + b = b + a$

associativity $(a + b) + c = a + (b + c)$

distributivity $(a + b)c = ac + bc$

unit element $1 \cdot a = a$

inverse element for all $a \neq 0$ exists the inverse element a^{-1} such that $aa^{-1} = 1$.

commutativity $ab = ba$

associativity $(ab)c = a(bc)$

---

[1] Many authors use the notation $R[X]$ with capital X. But capitals denote sets, too. Therefore I prefer the notation $R[x]$.

[2] To put $\deg 0 = -1$ seems to me no good choice neither.

A ring is called to have no null divisors if and only if $a \cdot b = 0$ implies that either $a = 0$ or $b = 0$. Such a ring is called an *integral domain*. A ring with no null divisors can be embedded into a quotient field. For any field $\mathbf{F}$, one can define the polynomial ring $\mathbf{F}[x]$. This is not a field, but a ring with important nice properties. Especially the following algorithms work:

- division with remainder;

- the Euclidean algorithm;

- the extended Euclidean algorithm.

- There exists a greatest common divisor for any two nonzero polynomials.

5.0.1 Lagrange's Theorem

Theorem 11 (Lagrange's Theorem). *Let R be an integral domain or F be any field. A polynomial $p(x) \in R[x]$ or $p(x) \in F[x]$ has at most as many roots as its degree d.*

Especially a polynomial $f(x) \in \mathbf{Z}_p[x]$, with any prime p, has at most as many roots as its degree.

Corollary 20. *Let p be a prime number. Assume for the monic polynomial $P \in \mathbf{Z}_p[x]$ there are known as many zeros $a_1 \ldots a_d$ as the degree $d = \deg P$ allows. Then*

$$(5.0.1) \qquad P(x) \equiv \prod_{1 \leq i \leq d} (x - a_i) \pmod{\mathbf{Z}_p}$$

Proof. Consider the difference

$$P(x) - \prod_{1 \leq i \leq d} (x - a_i) \in \mathbf{Z}_p[x]$$

Its degree is at most $d - 1$ but it has d zeros. This is impossible for a nonzero polynomial. Hence the difference is zero. $\square$

Problem 61. *Take the ring $\mathbf{Z}_9$ and the polynomial $x(x-3)(x^2-1)(x^2-4)$ as an instructive counterexample. Which important assumption about the ring is missing ? What is the degree. How many zeros does the polynomial have.*

Answer. This is an example of a polynomial of degree six with nine zeros. The ring $\mathbf{Z}_9$ is not an integral domain. The "too many" zeros $4, 5, 6$ occur because of two factors divisible by 3, whereas no factor is divisible by nine. □

Proposition 34. *The polynomial ring $\mathbf{Z}_m$ is an integral domain if an only if m is a prime number.*

5.1 Quadratic Residues

Definition 25 (Quadratic residue and nonresidue). Let p be a prime. The remainder class of a modulo p is called a *quadratic residue* if the congruence $x^2 \equiv a \pmod{p}$ is solvable.
The remainder class of a modulo p is called a *quadratic nonresidue* if the congruence $x^2 \equiv a \pmod{p}$ is not solvable.

Proposition 35 (Euler's criterium). *Let p be an odd prime and suppose that $p \nmid a$ is not a divisor of a. The quadratic congruence*

$$x^2 \equiv a \pmod{p} \quad \text{is solvable iff} \quad a^{\frac{p-1}{2}} \equiv 1 \pmod{p}$$

$$x^2 \equiv a \pmod{p} \quad \text{is not solvable iff} \quad a^{\frac{p-1}{2}} \equiv -1 \pmod{p}$$

Proof. Let
$$\mathbf{Z}_p^* = \{a \in \mathbf{Z}_p : p \nmid a\}$$
The mapping $x \in \mathbf{Z}_p^* \mapsto x^2 \in \mathbf{Z}_p^*$ is two to one. Indeed $x^2 \equiv y^2 \pmod{p}$ implies that either $x \equiv y \equiv 0 \pmod{p}$ which is excluded from the domain, or one gets two different solutions $x \equiv y \pmod{p}$ and $x \equiv -y \pmod{p}$. Hence the range

$$K = \{a \in \mathbf{Z}_p^* : x^2 \equiv a \pmod{p} \text{ is solvable}\}$$

has $\frac{p-1}{2}$ elements. Next we use Lagrange's Theorem 11. By Fermat's Little Theorem, the polynomial $x^{p-1} - 1$ has exactly $p - 1$ zeros, indeed these are the remainder classes $1, 2, \ldots p - 1$ modulo p. This turns out to be the maximal possible number of zeros. In the polynomial factoring

$$x^{p-1} - 1 = (x^{\frac{p-1}{2}} - 1)(x^{\frac{p-1}{2}} + 1)$$

each factor on the right-hand side has at most $\frac{p-1}{2}$ zeros but their product has $p - 1$ zeros. Hence the factors $x^{\frac{p-1}{2}} - 1$ and $x^{\frac{p-1}{2}} + 1$ have both exactly $\frac{p-1}{2}$ zeros. Thus the set

$$H = \{a \in \mathbf{Z}_p^* : a^{\frac{p-1}{2}} \equiv 1 \pmod{p}\}$$

5.1. QUADRATIC RESIDUES

turns out to have $\frac{p-1}{2}$ elements, too.

Assume the quadratic congruence $x^2 \equiv a \pmod{p}$ to be solvable. The assumption $p \nmid a$ implies $p \nmid x$. Hence Fermat's Little Theorem implies

$$a^{\frac{p-1}{2}} \equiv x^{p-1} \equiv 1 \pmod{p}$$

as claimed. In other words, we get the inclusion $K \subseteq H$. Since we know already that both sets K and H have the same number of elements, the equality $K = H$ occurs. Thus Euler's criterium is shown to be valid. □

Problem 62. *Give a simple proof of Euler's criterium assuming existence of a primitive root.*

Answer. Assume the quadratic congruence $x^2 \equiv a \pmod{p}$ to be solvable. The assumption $p \nmid a$ implies $p \nmid x$. Hence Fermat's Little Theorem implies

$$a^{\frac{p-1}{2}} \equiv x^{p-1} \equiv 1 \pmod{p}$$

as claimed.

Assume the quadratic congruence $x^2 \equiv a \pmod{p}$ not to be solvable. There exists s primitive root r modulo p. For this root holds

$$r^{\frac{p-1}{2}} \equiv -1 \pmod{p}$$

since the powers $\{r, r^2, \ldots r^{p-1}\}$ exhaust all $p-1$ congruence classes modulo p which are not divisible by p. The assumption $p \nmid a$ implies $a \equiv r^s$ to hold to some integer s. For even s, it is easy to see that $x = r^{s/2}$ solves the quadratic congruence. Hence s is odd and hence

$$a^{\frac{p-1}{2}} \equiv r^{\frac{s(p-1)}{2}} \equiv (-1)^s \equiv -1 \pmod{p}$$

□

Definition 26 (Legendre symbol). Let p be an odd prime and a any integer. The *Legendre symbol* is defined to be
(5.1.1)
$$\left(\frac{a}{p}\right) = \begin{cases} 1, & \text{if } x^2 \equiv a \pmod{p} \text{ is solvable and } p \nmid a; \\ -1, & \text{if } x^2 \equiv a \pmod{p} \text{ is not solvable and } p \nmid a; \\ 0, & \text{if } p \mid a. \end{cases}$$

Theorem 12 (The basic formula for the Legendre symbol).
Let p be an odd prime and a any integer. The Legendre symbol is given by Euler's formula

$$(5.1.2) \qquad \left(\frac{a}{p}\right) \equiv a^{\frac{p-1}{2}} \pmod{p}$$

Assume additionally that p does not divide a. Then the Legendre symbol, equals $+1$ if a is a quadratic residue, and -1 if a is a quadratic non-residue modulo p.

Problem 63. *Using Euler's criterium, we confirm by explicit calculation that for the case that $\frac{p^2-1}{8}$ is even, the congruence $x^2 \equiv 2 \pmod{p}$ has a solution. (It seems to be more difficult to confirm directly that for the case $\frac{p^2-1}{8}$ odd, the congruence $x^2 \equiv 2 \pmod{p}$ has no solution.)*

Answer. Let p be an odd prime. We check that for $\frac{p^2-1}{8}$ even, the congruence

$$x^2 \equiv 2 \pmod{p}$$

is solvable, by giving the solution. Let $x := 2^e$ with the exponent

$$(5.1.3) \qquad e = \begin{cases} \dfrac{p^2-1}{16} - \dfrac{p-5}{8} & \text{if } p \equiv 1 \pmod 8; \\ \dfrac{p^2-1}{16} - \dfrac{p-3}{4} & \text{if } p \equiv -1 \pmod 8 \end{cases}$$

The gentleman is calculating:

$$(5.1.4) \qquad 2e - 1 = \begin{cases} \dfrac{(p-1)^2}{8} & \text{if } p \equiv 1 \pmod 8; \\ \dfrac{(p-1)(p-3)}{8} & \text{if } p \equiv -1 \pmod 8 \end{cases}$$

and hence

$$x^2 = 2 \cdot \left(2^{p-1}\right)^{\frac{p-1}{8}} \equiv 2 \qquad \text{if } p \equiv 1 \pmod 8$$

$$x^2 = 2 \cdot \left(2^{\frac{p-1}{2}}\right)^{\frac{p-3}{4}} \equiv 2 \cdot (\pm 1)^{\frac{p-3}{4}} \equiv 2 \qquad \text{if } p \equiv -1 \pmod 8$$

In the first case, we use Fermat's Little Theorem $2^{p-1} \equiv 1$ directly, and obtain $x^2 \equiv 2$ as required. In the second case, we need only $2^{[(p-1)/2]} \equiv \pm 1$, but since $(p-3)/4$ is even, we get again $x^2 \equiv 2$ as required.

5.1. QUADRATIC RESIDUES

Problem 64. *Use Euler's criterium to show that all prime factors of $N = 1 + x^2$ are either 2 or primes $p \equiv 1 \pmod 4$.*

Answer. Assume that $p \mid N$ is a prime factor of $N = 1 + x^2$. Hence

$$x^2 \equiv -1 \pmod p$$

is solvable. Clearly $p \nmid x$. By Euler's criterium with $a = -1$ we get

$$(-1)^{\frac{p-1}{2}} = 1$$

Hence $(p-1)/2$ is even and $p \equiv 1 \pmod 4$.

5.1.1 Gauss' Proof of Quadratic Reciprocity

Let $m \geq 3$ be an odd integer. Let a be an integer such that $\gcd(a, m) = 1$. Let R be the set of remainders which occur for the divisions ia/m with $i = 1, 2, \ldots, \frac{m-1}{2}$. We partition the set R into the small remainders S and the large remainders L.

(5.1.5) $\quad R = \{r_i = ia - m\lfloor \frac{ia}{m} \rfloor \ : \ i = 1, 2, \ldots, \frac{m-1}{2}\}$
(5.1.6) $\quad S = \{r_i \in R : 0 < r_i < \frac{m}{2}\}$
(5.1.7) $\quad L = \{r_i \in R : \frac{m}{2} < r_i \leq m - 1\}$

Lemma 28 (Gauss' lemma). *The set R has $\frac{m-1}{2}$ elements. Moreover*

(5.1.8)
$$\begin{aligned} S \cup m - L &= \{r_i \in R : 0 < r_i < \tfrac{m}{2}\} \\ &\quad \cup \{m - r_i : r_i \in R \text{ and } \tfrac{m}{2} < r_i \leq m - 1\} \\ &= \{1, 2, \ldots, \tfrac{m-1}{2}\} \end{aligned}$$

In other word, the set $S \cup m - L$ is a permutation of the numbers $1, 2, \ldots, \frac{m-1}{2}$.

Proof. At first, we observe that for any $i, j = 1, 2, \ldots, \frac{m-1}{2}$

$$\begin{aligned} r_i + r_j &\equiv (i+j)a \pmod m \quad \text{and} \\ r_i - r_j &\equiv (i-j)a \pmod m \end{aligned}$$

Now $1 \leq (i+j) \leq m - 1$ implies that $i + j$ is not divisible by m. Similarly, for $i \neq j$ the difference is in the range $1 \leq |i - j| \leq m - 1$ and not divisible by m. Since $\gcd(a, m) = 1$ is assumed, we conclude that neither $(i+j)a$ nor $(i-j)a$ are the divisible by m.

Hence neither the sum $r_i + r_j$ nor the difference $r_i - r_j$ with $i \neq j$, are divisible by m.

Especially, we conclude that $r_i \neq r_j$ for $i \neq j$. Thus the set R has $\frac{m-1}{2}$ elements. Moreover, all elements of the set

$$S \cup m - L =$$
$$\{r_i \in R : 0 < r_i < \tfrac{m}{2}\} \cup \{m - r_i \in R : \tfrac{m}{2} < r_i \leq m - 1\}$$

appear only once. Hence this set has $\frac{m-1}{2}$ elements, too. Hence the claim (5.1.8) holds. □

Definition 27. I define the *parity* of any finite set to be $+1$ if it has an even number of members, and -1 if it has an odd number of elements. I define the *parity count* to be

(5.1.9)
$$(a/m) = \begin{cases} +1 \text{ if } L \text{ of } r_i > \tfrac{m}{2} \text{ has even number of elements;} \\ -1 \text{ if } L \text{ of } r_i > \tfrac{m}{2} \text{ has odd number of elements.} \end{cases}$$

In other words, the parity count is the parity of the set L of large remainders. Too, we may write

(5.1.10) $\qquad (a/m) = (-1)^{\#\{r_i > \frac{m}{2}\}} = (-1)^{\#\{i \,:\, \lfloor \frac{ia}{m} \rfloor > \frac{1}{2}\}}$

with $i = 1, 2, \ldots, \frac{m-1}{2}$.

Proposition 36. *Let $m, n \geq 3$ be odd integers. Let a be an integer such that $\gcd(a, m) = 1$. The parity count has the following properties*

(5.1.11) $\qquad (1/m) = 1$
(5.1.12) $\qquad (a/m) = (b/m) \quad \text{if } a \equiv b \pmod{m}$
(5.1.13) $\qquad (2a/m) = (2/m)(a/m)$
(5.1.14) $\qquad (2/m) = (-1)^{\frac{m^2-1}{8}}$
(5.1.15) $(n/m)(m/n) = (-1)^{\frac{n-1}{2} \frac{m-1}{2}}$

if $n, m \geq 3$ are odd and relatively prime.

Lemma 29. *With $r_i = ia - m \lfloor \frac{ia}{m} \rfloor$ for $i = 1, 2, \ldots \frac{m-1}{2}$ hold*

(5.1.16) $\quad \#\{r_i > \tfrac{m}{2}\} \equiv (a+1)\dfrac{m^2-1}{8}$
$\qquad + \sum \{\lfloor \tfrac{ia}{m} \rfloor : 1 \leq i \leq \tfrac{m-1}{2}\} \pmod{2} \quad \text{for all } a;$

5.1. QUADRATIC RESIDUES

(5.1.17) $$\#\{r_i > \tfrac{m}{2}\} \equiv \sum\{\lfloor \tfrac{ia}{m} \rfloor : 1 \le i \le \tfrac{m-1}{2}\} \pmod{2}$$
if a is odd.

Proof. Addition of the members of set $S \cup m - L$ from formula (5.1.8) yields

$$\sum\{r_i \in S\} + \sum\{m - r_i : r_i \in L\} = \sum\{i : 1 \le i \le \tfrac{m-1}{2}\}$$

(5.1.18) $$\sum\{r_i \in S\} + m|L| - \sum\{r_i \in L\} = \frac{m^2 - 1}{8}$$

Addition of the members of set $R = S \cup L$, yields

$$\sum\{r_i \in S\} + \sum\{r_i \in L\}$$
$$= a \sum\{i : 1 \le i \le \tfrac{m-1}{2}\} - m \sum\{\lfloor \tfrac{ia}{m} \rfloor : 1 \le i \le \tfrac{m-1}{2}\}$$

The first sum is easy calculated

(5.1.19)
$$\sum\{r_i \in S\} + \sum\{r_i \in L\}$$
$$= \frac{a(m^2 - 1)}{8} - m \sum\{\lfloor \tfrac{ia}{m} \rfloor : 1 \le i \le \tfrac{m-1}{2}\}$$

We add formulas (5.1.18) and (5.1.19). Finally use that m is assumed to be odd and obtain the first congruence (5.1.16).

$$2 \cdot \sum\{r_i \in S\} + m|L|$$
$$= (a+1)\frac{m^2 - 1}{8} - m \sum\{\lfloor \tfrac{ia}{m} \rfloor : 1 \le i \le \tfrac{m-1}{2}\}$$
$$|L| \equiv (a+1)\frac{m^2 - 1}{8} - \sum\{\lfloor \tfrac{ia}{m} \rfloor : 1 \le i \le \tfrac{m-1}{2}\} \pmod{2}$$

Since $(m^2 - 1)/8$ is an integer, one gets the second congruence (5.1.17) under the additional assumption that a is odd. □

Lemma 30.

(5.1.20) $$(2/m) = (-1)^{\frac{m^2-1}{8}}$$

More practically we get

$$(2/m) = \begin{cases} +1 & \text{if } m \equiv 1 \pmod{8} \text{ or } m \equiv 7 \pmod{8} \\ -1 & \text{if } m \equiv 3 \pmod{8} \text{ or } m \equiv 5 \pmod{8} \end{cases}$$

Proof. We put $a = 2$ into formula (5.1.16). Because of $\frac{2i}{m} < 1$ for $1 \leq i \leq \frac{m-1}{2}$ the sum vanishes and we obtain from definition (5.1.10)

$$\#\{r_i = 2i - m\lfloor \tfrac{2i}{m} \rfloor : r_i > \tfrac{m}{2}\} \equiv \tfrac{m^2-1}{8} + \sum\{\lfloor \tfrac{i}{m} \rfloor : 1 \leq i \leq \tfrac{m-1}{2}\}$$
$$\equiv \tfrac{m^2-1}{8} \pmod{2}$$

$$(2/m) = (-1)^{\frac{m^2-1}{8}}$$

$\square$

Lemma 31. *For odd integer a holds*

(5.1.21) $\qquad (a/m) = (-1)^{\sum\{\lfloor \frac{ai}{m} \rfloor : 1 \leq i \leq \frac{m-1}{2}\}}$

Proof. We use formulas (5.1.17) and (5.1.10)

$$(a/m) = (-1)^{\#\{r_i > \frac{m}{2}\}} = (-1)^{\sum\{\lfloor \frac{ia}{m} \rfloor : 1 \leq i \leq \frac{m-1}{2}\}}$$

$\square$

Lemma 32.

(5.1.22) $\qquad (a/m) = (-1)^{\sum\{\lfloor \frac{2ai}{m} \rfloor : 1 \leq i \leq \frac{m-1}{2}\}}$

holds for all integers a.

Proof. Elementary calculation gives

$$\lfloor \tfrac{2ai}{m} \rfloor = \begin{cases} 2\lfloor \tfrac{ai}{m} \rfloor & \text{if } \lfloor \tfrac{ai}{m} \rfloor < \tfrac{1}{2} \\ 2\lfloor \tfrac{ai}{m} \rfloor - 1 & \text{if } \lfloor \tfrac{ai}{m} \rfloor > \tfrac{1}{2} \end{cases}$$

and hence

$$\sum_{1 \leq i \leq \frac{m-1}{2}} \lfloor \tfrac{2ai}{m} \rfloor \equiv \#\{i : \lfloor \tfrac{ai}{m} \rfloor > \tfrac{1}{2}\} \pmod{2}$$

Now we use formula (5.1.10) to get the result.

(5.1.10) $\qquad (a/m) = (-1)^{\#\{i : \lfloor \frac{ai}{m} \rfloor > \frac{1}{2}\}}$

$\qquad\qquad (a/m) = (-1)^{\sum\{\lfloor \frac{2ai}{m} \rfloor : 1 \leq i \leq \frac{m-1}{2}\}}$

$\square$

5.1. QUADRATIC RESIDUES

Lemma 33.

(5.1.23) $\qquad (2a/m) = (2/m)(a/m)$

Proof. In formula (5.1.16) we replace a by $2a$ to get the following congruences modulo 2

$$\#\{r_i = 2ai - m\lfloor \tfrac{2ai}{m}\rfloor : r_i > \tfrac{m}{2}\}$$
$$\equiv (2a+1)\tfrac{m^2-1}{8} + \sum\{\lfloor \tfrac{2ai}{m}\rfloor : 1 \le i \le \tfrac{m-1}{2}\} \pmod{2}$$
$$\#\{r_i = 2ai - m\lfloor \tfrac{2ai}{m}\rfloor : r_i > \tfrac{m}{2}\}$$
$$\equiv \tfrac{m^2-1}{8} + \sum\{\lfloor \tfrac{2ai}{m}\rfloor : 1 \le i \le \tfrac{m-1}{2}\} \pmod{2}$$

On the left-hand side we use definition (5.1.10) and replace a by $2a$. On the right-hand side we use formula (5.1.22) which holds for <u>all</u> integers a.

$$(2a/m) = (-1)^{\tfrac{m^2-1}{8}} \cdot (-1)^{\sum\{\lfloor \tfrac{2ai}{m}\rfloor : 1 \le i \le \tfrac{m-1}{2}\}}$$
$$(2a/m) = (2/m)(a/m)$$

$\square$

Lemma 34.

$$(n/m)(m/n) = (-1)^{\tfrac{n-1}{2}\tfrac{m-1}{2}}$$

if $n, m \ge 3$ are odd and relatively prime.

Proof. By a *grid point* is meant a point (x,y) with both coordinates integers.

$$R = \{(x,y) : 0 < x < \tfrac{m}{2} \text{ and } 0 < y < \tfrac{n}{2}\}$$
$$S = \{(x,y) : 0 < x < \tfrac{m}{2}, 0 < y < \tfrac{n}{2} \text{ and } nx < my\}$$
$$D = \{(x,y) : 0 < x < \tfrac{m}{2}, 0 < y < \tfrac{n}{2} \text{ and } nx = my\}$$
$$T = \{(x,y) : 0 < x < \tfrac{m}{2}, 0 < y < \tfrac{n}{2} \text{ and } my < nx\}$$

The open rectangle R is partitioned into the upper triangle S, the diagonal D and the lower triangle T. We count the grid points in the upper triangle by adding horizontal strips and get

$$\sum\{\lfloor \tfrac{my}{n}\rfloor : 1 \le y \le \tfrac{n-1}{2}\} \text{ grid points in } S.$$

We now count the grid points in the lower triangle by adding vertical strips and get

$$\sum\{\lfloor \tfrac{nx}{m}\rfloor : 1 \le x \le \tfrac{m-1}{2}\} \text{ grid points in } T.$$

Clearly the open rectangle R contains $\frac{m-1}{2} \cdot \frac{n-1}{2}$ grid points. Since m and n are relatively prime, there are no grid points on the diagonal D. Hence

$$\frac{m-1}{2} \cdot \frac{n-1}{2}$$
$$= \sum\{\lfloor \tfrac{my}{n} \rfloor : 1 \leq y \leq \tfrac{n-1}{2}\} + \sum\{\lfloor \tfrac{nx}{m} \rfloor : 1 \leq x \leq \tfrac{m-1}{2}\}$$

We use formula

(5.1.21) $\qquad (a/m) = (-1)^{\sum\{\lfloor \frac{ai}{m} \rfloor : 1 \leq i \leq \frac{m-1}{2}\}}$

twice and get

$$(-1)^{\frac{m-1}{2}\frac{n-1}{2}} = (-1)^{\sum\{\lfloor \frac{my}{n} \rfloor : 1 \leq y \leq \frac{n-1}{2}\}} \cdot (-1)^{\sum\{\lfloor \frac{nx}{m} \rfloor : 1 \leq x \leq \frac{m-1}{2}\}}$$
$$= (m/n)(n/m)$$

$\square$

Lemma 35. *Let $m \geq 3$ be an odd integer. Let a be an integer such that $\gcd(a, m) = 1$.*

(5.1.24) $\qquad (\tfrac{m-1}{2})! \, (a/m) \equiv (\tfrac{m-1}{2})! \, a^{\frac{m-1}{2}} \pmod{m}$

Proof. Multiply the members of set $S \cup m - L$ from formula (5.1.8). Formula (5.1.10) yields

$$\prod\{r_i \in S\} \cdot \prod\{m - r_i : r_i \in L\} = (\tfrac{m-1}{2})!$$
$$\prod\{r_i \in S\} \cdot \prod\{r_i \in L\} \cdot (-1)^{\#\{r_i > \frac{m}{2}\}} \equiv (\tfrac{m-1}{2})! \pmod{m}$$
(5.1.25) $\qquad \prod\{r_i \in S\} \cdot \prod\{r_i \in L\} \equiv (\tfrac{m-1}{2})! \, (a/m) \pmod{m}$

The remainders $r_i = ia - m\lfloor \tfrac{ia}{m} \rfloor$ for $i = 1, 2, \ldots, \tfrac{m-1}{2}$ are multiplied to obtain

$$\prod\{r_i \in S\} \cdot \prod\{r_i \in L\} \equiv \prod\{ia : 1 \leq i \leq \tfrac{m-1}{2}\} \pmod{m}$$
(5.1.26)
$$\prod\{r_i \in S\} \cdot \prod\{r_i \in L\} \equiv (\tfrac{m-1}{2})! \, a^{\frac{m-1}{2}} \pmod{m}$$

Both formulas (5.1.25) and (5.1.26) together yield the result. $\square$

Proposition 37. *Let $p \geq 3$ be an odd prime. Let a be an integer not divisible by the prime p.*

(5.1.27) $\qquad (a/p) \equiv a^{\frac{p-1}{2}} \pmod{p}$

5.1. QUADRATIC RESIDUES

Proof. Since the prime p does not divide $(\frac{p-1}{2})!$ the congruence (5.1.27) follows from the congruence (5.1.24). □

Problem 65. *Prove that for all odd composite numbers $m \geq 15$, the factorial $(\frac{m-1}{2})!$ is divisible by m. Hint: You may use Legendre's proposition 13.*

Solution of problem 65. Let m be odd and composite. I distinguish two cases:

m is a prime power and $m \neq 9$. Assume $m = p^s$ and let $p^r \| (\frac{m-1}{2})!$ By Legendre's proposition (13) holds $r \geq \lfloor \frac{m-1}{2p} \rfloor$. We check that for $p = 3$ and $s \geq 3$ as well as for $p \geq 5$ and $s \geq 2$ holds $2s < p^{s-1}$. Hence

$$s \leq \lfloor \frac{p^s - 1}{2p} \rfloor \leq \lfloor \frac{m-1}{2p} \rfloor \leq r$$

confirming that $s \leq r$ and hence $m \mid (\frac{m-1}{2})!$ is proved.

m is divisible by two different primes. Take any odd prime p. Let $p^s \| m$ be the highest prime power dividing m and let $p^r \| (\frac{m-1}{2})!$ There exists an odd prime $q \neq p$ for which holds $p^s q \mid m$. We check $2s < p^{s-1} q$. By Legendre's proposition (13) holds $r \geq \lfloor \frac{m-1}{2p} \rfloor$. Hence

$$s \leq \lfloor \frac{p^s \cdot q - 1}{2p} \rfloor \leq \lfloor \frac{m-1}{2p} \rfloor \leq r$$

confirming that $s \leq r$.

Since $p^s \| m \Rightarrow p^s \mid (\frac{m-1}{2})!$ holds for all odd primes p, one concludes $m \mid (\frac{m-1}{2})!$ as claimed.

Thus we have covered all composite odd numbers except $m = 9$. One check directly that $m \nmid (\frac{m-1}{2})!$ for $m = 9$. □

Problem 66. *Hence the congruence (5.1.24) gives some extra information only for $m = 9$,—except the really important case that m is prime. What may one claim?*

Answer. Let $m = 9$ and let a be an integer not divisible by 3. Since 3 but not 9 divides $(\frac{m-1}{2})! = 4! = 24$, one gets from congruence (5.1.24)

$$(a/9) = 1 \equiv a^4 \pmod{3} \quad \text{for } a = 1, 2, 4, 5, 7, 8$$

which is clearly true. [3] □

---
[3] Better trivial than wrong

Lemma 36. *Let $p \geq 3$ be an odd prime. Let a be an integer not divisible by the prime p.*

(5.1.28)
$$\left(\frac{a}{p}\right) = (a/p)$$

The parity count is an extension of the Legendre symbol.

Proof. Formula (5.1.2) for the Legendre symbol from theorem 12, and lemma 37 yield

$$\left(\frac{a}{p}\right) \equiv a^{\frac{p-1}{2}} \pmod{p}$$

$$a^{\frac{p-1}{2}} \equiv (a/p) \pmod{p}$$

$$\left(\frac{a}{p}\right) = (a/p) \quad \text{since they assume only the values } \pm 1.$$

□

Now lemma 34 and formula (5.1.20) yield Gauss' main result.

Theorem 13 (Quadratic reciprocity, Gauss 1795). *For any odd primes $p \neq q$*

(5.1.29)
$$\left(\frac{p}{q}\right)\left(\frac{q}{p}\right) = (-1)^{\frac{p-1}{2} \frac{q-1}{2}}$$

(5.1.30)
$$\left(\frac{2}{p}\right) = (-1)^{\frac{p^2-1}{8}}$$

Problem 67. *Use quadratic reciprocity to show that all prime factors of $N = 1 + a + a^2$ are either 3 or primes $p \equiv 1 \pmod{3}$.*

Answer. Assume that $p \mid N$ is a prime factor of $N = 1 + a + a^2$. Complete the square to get $(2a+1)^2 + 3 = 4N$ and put $x := 2a+1$. The quadratic congruence

$$x^2 \equiv -3 \pmod{p}$$

is solvable. In other words, -3 is a quadratic residue modulo p. We now need to assume $p \neq 3$. Since $3 \nmid p$ we get Legendre symbol $+1$ and use quadratic reciprocity to go on:

$$1 = \left(\frac{-3}{p}\right) = \left(\frac{-1}{p}\right) \cdot \left(\frac{3}{p}\right) = (-1)^{\frac{p-1}{2}} \cdot (-1)^{\frac{p-1}{2} \frac{3-1}{2}} \left(\frac{p \bmod 3}{3}\right)$$

$$= \left(\frac{p \bmod 3}{3}\right)$$

Hence $p \equiv 1 \bmod 3$ for all prime factors except $p = 3$. It is easy to see that $3 \mid 1 + a + a^2$ if and only $a \equiv 1 \pmod{3}$.

5.1.2 Quadratic Residues for Composite Numbers

Theorem 14. *Let $m > 1$ have the prime factorization*

$$m = 2^r \prod_{1 \le i \le t} p_i^{r_i}$$

and assume $\gcd(m, a) = 1$. Then a is a quadratic residue modulo m if and only if

(a) *For all i the Legendre symbol is $\left(\frac{a}{p_i}\right) = 1$;*

(b1) *If $4 \mid m$ and $8 \nmid m$ then $a \equiv 1 \mod 4$;*

(b2) *If $8 \mid m$ then $a \equiv 1 \mod 8$.*

Problem 68. *Find all possible cases for the last two digits that an odd perfect square may have.*

Solution. Take the case that $5 \nmid a$. By theorem 14 the quadratic congruence $x^2 \equiv a \pmod{100}$ is solvable if and only if

(a) the Legendre symbol is $\left(\frac{a}{5}\right) = 1$;

(b1) $a \equiv 1 \mod 4$;

The condition (a) holds if and only if either $a \equiv 1 \mod 5$ or $a \equiv 4 \mod 5$. Both conditions (a) and (b1) hold if and only if either $a \equiv 1 \mod 20$ or $a \equiv 9 \mod 20$. Take the case that $5 \mid a$. The quadratic congruence $x^2 \equiv a \pmod{100}$ is equivalent to $y^2 \equiv b \pmod 4$ with $x = 5y$ and $a = 25b$. Hence $25 \mid a$. Since $\gcd(b, 4) = 1$ the latter congruence solvable if and only if $b \equiv 1 \mod 4$ hence if and only if $a \equiv 25 \mod 100$.

Altogether we have obtained as possible cases for the last two digits

$$01,\ 21,\ 41,\ 61,\ 81,\ 09,\ 29,\ 49,\ 69,\ 89,\ 25$$

□

Problem 69. *Find all possible cases for the last two digits that an even perfect square may have.*

Solution. The quadratic congruence $x^2 \equiv a \pmod{100}$ is equivalent to $y^2 \equiv b \pmod{25}$ with $x = 2y$ and $a = 4b$. Hence $4 \mid a$. Take the case that $5 \nmid a$. By theorem 14 the quadratic congruence $y^2 \equiv b \pmod{25}$ is solvable if and only if the Legendre symbol is $\left(\frac{b}{5}\right) = 1$. This condition holds if and only if either $b \equiv 1 \mod 5$ or $b \equiv 4 \mod 5$. Finally one concludes that the quadratic congruence $x^2 \equiv a \pmod{100}$ with $5 \nmid a$ is solvable if and only if either $a \equiv 4 \mod 20$ or $a \equiv 16 \mod 20$.

Take the case that $5 \mid a$. Hence $10 \mid a$ and $100 \mid a$. We get only the case $a \equiv 0 \pmod{100}$. Altogether we have obtained as possible cases for the last two digits

$$04,\, 24,\, 44,\, 64,\, 84,\, 16,\, 36,\, 56,\, 76,\, 96,\, 00$$

$\square$

Problem 70. *Give a constructive direct proof of theorem 14.*

Lemma 37. *Assume the odd prime $p \nmid a$ does not divide $a \geq 1$. Let $s \geq 0$. If the quadratic congruence $x^2 \equiv a \pmod{p^{2^s}}$ is solvable then the quadratic congruence $x^2 \equiv a \pmod{p^{2 \cdot 2^s}}$ is solvable.*

Proof. Assume that the quadratic congruence $x^2 \equiv a \mod p^{2^s}$ holds. We put $z = x + p^{2^s} y$ to solve

(5.1.31)
$$z^2 \equiv a \pmod{p^{2 \cdot 2^s}}$$
$$x^2 + 2p^{2^s} xy + p^{2 \cdot 2^s} y^2 \equiv a \pmod{p^{2 \cdot 2^s}}$$
$$2p^{2^s} x^2 y \equiv x(a - x^2) \pmod{p^{2 \cdot 2^s}}$$
$$2x^2 y \equiv x \cdot \frac{a - x^2}{p^{2^s}} \pmod{p^{2^s}}$$
$$2ay \equiv x \cdot \frac{a - x^2}{p^{2^s}} \pmod{p^{2^s}}$$
$$y \equiv b \cdot x \cdot \frac{a - x^2}{p^{2^s}} \pmod{p^{2^s}}$$

Since $p \nmid 2a$ there exists an inverse b such that $2ab \equiv 1 \mod p^{2^s}$. The number y has to be chosen according to the last line. All steps of the calculation are equivalences. We have indeed as required solved the congruence (5.1.31). $\square$

Lemma 38. *Assume the odd prime $p \nmid a$ does not divide $a \geq 1$. Let $s \geq 0$. If the quadratic congruence $x^2 \equiv a \pmod{p^{2^s}}$ is solvable then the quadratic congruence $x^2 \equiv a \pmod{2 \cdot p^{2^s}}$ is solvable.*

5.1. QUADRATIC RESIDUES

Proof. Assume that the quadratic congruence $x^2 \equiv a \pmod{p^{2^s}}$ holds. If x is even the congruence $x^2 \equiv a \pmod{2 \cdot p^{2^s}}$ holds, too, and we are ready. If x is odd we put $z = x + p^{2^s}$ which is even and get a solution of the congruence $z^2 \equiv a \pmod{2 \cdot p^{2^s}}$ as required. □

Lemma 39. *Assume the odd prime $p \nmid a$ does not divide $a \geq 1$. Let $r \geq 1$. The quadratic congruence $x^2 \equiv a \mod 2p^r$ is solvable if and only if the The quadratic congruence $x^2 \equiv a \mod p$ is solvable.*

Proof. Assume the quadratic congruence $x^2 \equiv a \mod p$ is solvable. We use lemma 37 to show by induction that for all s the quadratic congruence $x^2 \equiv a \pmod{p^{2^s}}$ is solvable. By lemma 38 the quadratic congruence $x^2 \equiv a \pmod{2 \cdot p^{2^s}}$ is solvable, too. Given any $r \geq 1$, there exists s such that $r \leq 2^s$. Hence the quadratic congruence $x^2 \equiv a \mod 2p^r$ is solvable. Conversely the last assertion implies that the quadratic congruence $x^2 \equiv a \mod p$ is solvable. □

Lemma 40. *Assume that $a \geq 1$ is odd. Let $t \geq 3$. If the quadratic congruence $x^2 \equiv a \pmod{2^t}$ is solvable then the quadratic congruence $x^2 \equiv a \pmod{2^{2t-2}}$ is solvable.*

Proof. Assume that the quadratic congruence $x^2 \equiv a \pmod{2^t}$ holds.
We put $z = x + 2^{t-1} y$ to solve

$$(5.1.32) \qquad z^2 \equiv a \pmod{2^{2t-2}}$$
$$x^2 + 2^t xy + 2^{2(t-1)} y^2 \equiv a \pmod{2^{2(t-1)}}$$
$$2^t x^2 y \equiv x(a - x^2) \pmod{2^{2(t-1)}}$$
$$ay \equiv x \cdot \frac{a - x^2}{2^t} \pmod{2^{t-2}}$$
$$y \equiv b \cdot x \cdot \frac{a - x^2}{2^t} \pmod{2^{t-2}}$$

Since a is odd there exists an inverse b such that $ab \equiv 1 \mod 2^{t-2}$. The number y has to be chosen according to the last line. All steps of the calculation are equivalences. We have indeed as required solved the congruence (5.1.32). □

Lemma 41. *Assume that $a \geq 1$ is odd. The quadratic congruence $x^2 \equiv a \pmod 8$ is solvable if and only if $a = 1$. For any $t \geq 3$, the quadratic congruence $x^2 \equiv a \pmod{2^t}$ is solvable if and only if $a \equiv 1 \pmod 8$.*

Proof. Assume that $a \equiv 1 \pmod 8$. It is obvious that the quadratic congruence $x^2 \equiv a \pmod 8$ is solvable. We use lemma 40 to show by induction that for all $2^t \geq 8$ the quadratic congruence $x^2 \equiv a \pmod{2^t}$ is solvable. The induction steps leads from modulus 2^t to 2^{2t-2}. Thus one needs to assume $2t - 2 > t$ and hence $t \geq 3$.

Conversely the last assertion implies that the quadratic congruence $x^2 \equiv a \mod 8$ is solvable. Since $x^2 \equiv 1 \pmod 8$ holds for all odd numbers x, we conclude that $a \equiv 1 \pmod 8$. □

A constructive proof of theorem 14. Proof of theorem 14 is done by induction on the number t of different primes in m.

Begin with the case $t = 1$. Here $m = p^r$ is a prime power. In the case that p is an odd prime the assertion is shown by lemma 39. The special case $p = 2$ with $2^r \geq 8$ is covered by lemma 40. The reader may check that the assertion holds for cases $m = 2$ and $m = 4$, too.

We proceed now by induction on the number of different primes. Assume the assertion is true for any m with less than t different prime factors. Let mn have t different prime factors but both m and n less than t prime factors. Too, we assume $\gcd(m, n) = 1$. Take any a such that $\gcd(a, mn) = 1$. Since $\gcd(a, m) = \gcd(a, n) = 1$ we use the induction assumption to solve the quadratic congruences

$$y^2 \equiv a \pmod m$$
$$z^2 \equiv a \pmod n$$

Next we use the Chinese remainder Theorem to get x such that

$$x \equiv y \pmod m$$
$$x \equiv z \pmod n$$

Together we have obtained a solution of

$$x^2 \equiv y^2 \equiv a \pmod m$$
$$x^2 \equiv z^2 \equiv a \pmod n$$

and hence $x^2 \equiv a \pmod{mn}$ as claimed.

There is a modification in the special case that m has a sing le prime factor $p = 2$. This factor has to be joined to any odd prime power p^r. Invoking lemma 39 we solve the congruence $z^2 \equiv a \pmod{2 \cdot p^r}$. □

5.2 Jacobi Symbols

Definition 28 (J-symbol). The *J-symbol* $\left(\frac{a}{m}\right)$ is defined as a natural extension of the parity count. For arbitrary top integer a and the odd bottom integer $m \geq 3$ we define

$$(5.2.1) \qquad \left(\frac{a}{m}\right) = \begin{cases} (a/m) & \text{if } \gcd(a, m) = 1 \\ 0 & \text{if } \gcd(a, m) > 1 \end{cases}$$

Corollary 21. *The J-symbol is an extension of the Legendre symbol.*

Proof. Because of formula (5.1.2) and proposition 37 from below, the parity count is an extension of the Legendre symbol $\left(\frac{a}{p}\right)$ for the case $p \nmid a$. By definition, the J-symbol is the extension of the parity count including the case $p \mid a$ with zero value. Similarly one gets $\left(\frac{a}{p}\right) = 0$ for the case $p \mid a$ from formula (5.1.2). Therefore the J-symbol remains an extension of the Legendre symbol, for both cases: $p \nmid a$ as well as $p \mid a$. This justifies equal notations for the J-symbol and the Legendre symbol. □

Remark. The symbols (a/m) respectively $\left(\frac{a}{m}\right)$ are only defined for odd $m \geq 3$. One might be tempted to put $\left(\frac{a}{1}\right) = 1$ for $a \neq 0$. Unless one restricts the modular calculations to $m \geq 3$, one is lead to the contradiction $\left(\frac{a}{1}\right) = \left(\frac{a \bmod 1}{1}\right) = \left(\frac{0}{1}\right) = 0$. For this reason we have assumed the bottom integer to be $m \geq 3$.

5.2.1 Calculation of J-symbols

An effective calculation of the parity count and indeed the J-symbol, too, is done by an algorithm analogous to the Euclidean algorithm. One uses once more a sequence of successive remainders. New features are factoring out powers of two and using quadratic reciprocity.

Algorithm 1 (Jacobi-type Eucliclean algorithm). *Let $m \geq 3$ be an odd integer and let a be any integer. The algorithm calculates the symbols (a/m) respectively $\left(\frac{a}{m}\right)$.*

division with remainder: *If $a > m$, we use division $a = qm + b$ with remainder $b = (a \bmod m)$. Now holds $(a/m) = (b/m)$.*

special case $m \mid a$: *If $m \mid a$ we obtain the greatest common divisor to be $\gcd(a,m) = m > 1$. Moreover, the J-symbol. is $\left(\frac{a}{a}\right) = \left(\frac{0}{a}\right) = 0$. The algorithm stops.* [4]

elimination of factors $4 \mid m$: *If $4 \mid m$, we use the property $(a/m) = (\frac{a}{4}/m)$.*

elimination of a last factor $2 \mid m$: *If $4 \nmid m$ but $2 \mid m$ we use the properties $(a/m) = (2/m)(\frac{a}{2}/m)$ together with*

$$(2/m) = \begin{cases} +1 & \text{if } m \equiv 1 \pmod{8} \text{ or } m \equiv 7 \pmod{8} \\ -1 & \text{if } m \equiv 3 \pmod{8} \text{ or } m \equiv 5 \pmod{8} \end{cases}$$

stopping condition $a = 1$: *If $a = 1$ but $m \geq 3$ we use the property $(1/m) = 1$ and the algorithm stops.*

otherwise $a \geq 3$ **is odd** *and we use quadratic reciprocity*

$$(a/m) = \begin{cases} -(m/a) & \text{if } a \equiv 3 \pmod{4} \text{ and } m \equiv 3 \pmod{4} \\ +(m/a) & \text{otherwise} \end{cases}$$

Next one goes back to the first case.

The algorithm stops if either $\pm(1/m)$ or $(0/m)$ is reached.

Remark. For the common case $\gcd(a,m) = 1$, the parity count (a/m) is uniquely calculated from its properties stated in proposition 36. Even more, the case $\gcd(a,m) = g > 1$, is included in the algorithm. Here the (extended) J-symbol. $\left(\frac{a}{m}\right)$ turns out to be zero. Too, one gets the greatest common divisor by the Euclidean algorithm as usual. Indeed, $\left(\frac{a}{m}\right) = 0$ if and only if $\gcd(a,m) > 1$.

Proposition 38. *As consequences of algorithm 1 one gets the following facts:*

(i) *The parity count and the J-symbol can be calculated in time proportional to the logarithm of the numbers involved.*

(ii) *The parity count is uniquely determined by its properties gathered in proposition 36.*

(iii) *The J-symbol takes only the values $0, 1, -1$.*

(iv) $\left(\frac{a}{m}\right) = 0$ *if and only if* $\gcd(a,m) > 1$.

---
[4]In this case, the parity count is undefined.

5.2. JACOBI SYMBOLS

Proposition 39 (Uniqueness of the J-symbol). *The J-symbol is <u>uniquely</u> determined from the following properties:*

(5.2.2) $$\left(\frac{0}{m}\right) = 0$$

(5.2.3) $$\left(\frac{2a}{m}\right) = \left(\frac{2}{m}\right) \cdot \left(\frac{a}{m}\right)$$

(5.2.4) $$\left(\frac{a}{m}\right) = \left(\frac{a \bmod m}{m}\right)$$

(5.2.5) $$\left(\frac{2}{m}\right) = (-1)^{\frac{m^2-1}{8}}$$

(5.2.6) $$\left(\frac{n}{m}\right) = (-1)^{\frac{n-1}{2}\frac{m-1}{2}} \left(\frac{m}{n}\right)$$

Corollary 22. *Especially the properties of the parity count gathered in proposition 36 already imply that for $m = p$ odd prime as denominator, the J-symbol equals the Legendre symbol $\left(\frac{a}{p}\right)$ from Euler's basic formula*

(5.1.2) $$\left(\frac{a}{p}\right) \equiv a^{\frac{p-1}{2}} \pmod{p}$$

Proof. One only needs still the lemma 42 below. Thus the stated properties imply the the algorithm 1 stops in all cases and leads to a unique calculation of the J-symbol. □

Lemma 42.

(5.2.7) $$\left(\frac{1}{m}\right) = 1$$

for any odd integer $m \geq 3$.

Proof. From the assumed formulas (5.2.3) and (5.2.5) we get

$$\left(\frac{2}{m}\right) = \left(\frac{2 \cdot 1}{m}\right) = \left(\frac{2}{m}\right) \cdot \left(\frac{1}{m}\right) \quad \text{and}$$

$$\left(\frac{2}{m}\right) = (-1)^{\frac{m^2-1}{8}} \neq 0 \quad \text{which imply}$$

$$\left(\frac{1}{m}\right) = 1 \quad \text{as claimed.}$$

□

Lemma 43. *The J–symbol is zero* $\left(\frac{a}{m}\right) = 0$ *if and only if the numbers a and m are not relatively prime.*

Proof. For $g = \gcd(a, m) > 1$, we have defined $\left(\frac{a}{m}\right) = 0$.

Conversely, assume $\left(\frac{a}{m}\right) = 0$. Since m is odd, the common divisor $g = \gcd(a, m)$, is odd. The algorithm to calculate the J-symbol produces in all steps symbols where top and bottom are both divisible by g. It stops with either top integer either 0 or 1. The second possibility has been excluded by the assumption $\left(\frac{a}{m}\right) = 0$.

At stop the top integer is zero and bottom integer is greater equal to 3. Hence $g \geq 3$. □

Lemma 44.

(5.2.8) $$\left(\frac{-1}{m}\right) = (-1)^{\frac{m-1}{2}}$$

for any odd integer $m \geq 3$.

Proof. Let $m = 1 + 2^r k$ where k is odd. In the case that $m \equiv 3$ (mod 4), we have $r = 1$ and $m - 1 = 2k$.

$$\left(\frac{-1}{m}\right) = \left(\frac{m-1}{m}\right) = \left(\frac{2}{m}\right) \cdot \left(\frac{k}{m}\right)$$
$$= (-1)^{\frac{m^2-1}{8}} \cdot (-1)^{\frac{m-1}{2} \frac{k-1}{2}} \left(\frac{m}{k}\right)$$
$$= (-1)^{\frac{m-1}{2} \frac{(m+1)/2 + k - 1}{2}} \left(\frac{1+2k}{k}\right)$$
$$= (-1)^{\frac{m-1}{2}} = -1$$

since $(m+1)/4 + (k-1)/2 = (2k+2)/4 + (k-1)/2 = k$ is odd. In the case that $m \equiv 1$ (mod 4), we get $r \geq 2$ and hence

$$\left(\frac{-1}{m}\right) = \left(\frac{m-1}{m}\right) = \left(\frac{2}{m}\right)^r \cdot \left(\frac{k}{m}\right)$$
$$= (-1)^{\frac{m-1}{2} \frac{r(m+1)/2 + k - 1}{2}} \left(\frac{1+2^r k}{k}\right)$$
$$= (-1)^{\frac{m-1}{2}} \cdot 1 = 1 = (-1)^{\frac{m-1}{2}}$$

For $r = 2$, we see $\frac{m-1}{2}$ is even and $r(m+1)/4 + (k-1)/2 = 2(2k+2)/4 + (k-1)/2$ is an integer. For $r \geq 3$, we see $\frac{m-1}{2} = 2^{r-1}k$ is divisible by 4 and $r(m+1)/4 + (k-1)/2 = r(2k+2)/4 + (k-1)/2$ is half of an integer. Both cases give an even exponent of -1 and confirm the result $\left(\frac{-1}{m}\right) = 1$. □

5.2.2 Jacobi Symbols

The extension of the Legendre symbol by means of the parity count depends on the detailed review of Gauss' proof of quadratic reciprocity given above. I do not know (and cannot find out exactly) whether and where this,—a bid strange—approach to the J-symbol appeared first or if it is in fact new. In the end comes as a relief the more classical and natural construction of the Jacobi which is explained and discussed below. It remains still surprising that the so differently constructed J-symbol and Jacobi symbol turn out to be synonymous.

Definition 29 (Jacobi symbols). The *Jacobi symbol* is defined for arbitrary top integer a, and any <u>positive</u> and <u>odd</u> bottom integer $m \geq 3$. Let

$$(5.2.9) \qquad m = \prod p_i^{r_i}$$

be its prime factorization. We define the *Jacobi symbol* by setting

$$(5.2.10) \qquad \left(\frac{a}{m}\right) = \prod \left(\frac{a}{p_i}\right)^{r_i}$$

with Legendre symbols on the right-hand side.

Lemma 45. *For any odd integers a, b, c, d holds*

$$(5.2.11) \qquad \frac{(ab)^2 - 1}{8} \equiv \frac{a^2 - 1}{8} + \frac{b^2 - 1}{8} \pmod{8}$$

$$(5.2.12) \qquad \frac{ab - 1}{2} \equiv \frac{a - 1}{2} + \frac{b - 1}{2} \pmod{2}$$

$$(5.2.13) \qquad \frac{ab-1}{2} \cdot \frac{cd-1}{2} \equiv \left[\frac{a-1}{2} + \frac{b-1}{2}\right] \cdot \left[\frac{c-1}{2} + \frac{d-1}{2}\right] \pmod{2}$$

Reason. For $a = 2k + 1$ odd the identity

$$a^2 = 4k^2 + 4k + 1 = 8\frac{(k+1)k}{2} + 1$$

shows that $a^2 - 1$ is divisible by 8. Hence

$$\frac{(ab)^2 - 1}{8} - \frac{a^2 - 1}{8} - \frac{b^2 - 1}{8} = \frac{(a^2 - 1)(b^2 - 1)}{8}$$

is divisible by 8, confirming formula (5.2.11).

$$\frac{ab-1}{2} = \frac{(a-1)(b-1)+a+b-2}{2}$$
$$= 2 \cdot \frac{a-1}{2} \cdot \frac{b-1}{2} + \frac{a-1}{2} + \frac{b-1}{2}$$
$$\frac{ab-1}{2} \equiv \frac{a-1}{2} + \frac{b-1}{2} \pmod{2}$$
$$\frac{cd-1}{2} \equiv \frac{c-1}{2} + \frac{d-1}{2} \pmod{2}$$

Multiplying yields the results (5.2.12) and (5.2.13). $\square$

Proposition 40. *The Jacobi symbol is an extension of the Legendre symbol and has the following properties:*

(5.2.14) $\qquad \left(\dfrac{0}{m}\right) = 0$

(5.2.15) $\qquad \left(\dfrac{a}{mn}\right) = \left(\dfrac{a}{m}\right) \cdot \left(\dfrac{a}{n}\right)$

(5.2.16) $\qquad \left(\dfrac{ab}{m}\right) = \left(\dfrac{a}{m}\right) \cdot \left(\dfrac{b}{m}\right)$

(5.2.17) $\qquad \left(\dfrac{a}{m}\right) = \left(\dfrac{a \bmod m}{m}\right)$

(5.2.18) $\qquad \left(\dfrac{2}{m}\right) = (-1)^{\frac{m^2-1}{8}}$

(5.2.19) $\qquad \left(\dfrac{n}{m}\right) = (-1)^{\frac{n-1}{2} \frac{m-1}{2}} \left(\dfrac{m}{n}\right)$

for any integers a, b and $n, m \geq 3$ any <u>odd</u> integers.

Corollary 23. *The J-symbol and the Jacobi symbol are equal, and have equal domain of definition.*

Proof. The properties of the Jacobi symbol gathered in proposition 40 include the properties of the J-symbol gathered in proposition 39. The latter (weaker) properties already imply the uniqueness of the J-symbol. Since the J-symbol and the Jacobi symbol have equal domains of definition, too, uniqueness implies their equality. $\square$

Remark. This corollary is astonishing.

5.2. JACOBI SYMBOLS

Proof of proposition 40. The first formula follows directly from the definition (5.2.10) of the Jacobi symbol. The remaining formulas are proved by induction on the number of prime factors in the denominator.

The next two formulas are to be obtained from the corresponding formula for the Legendre symbol, which in turn follow from Euler's formula (5.1.2).

The two last formulas are known to hold for the Legendre symbol by quadratic reciprocity, Theorem 13. For an inductive proof, say on the number of prime factors, let us suppose that the formula holds with m and n. Hence the formula (5.2.11) from lemma 45 yields the result:

$$\left(\frac{2}{mn}\right) = \left(\frac{2}{m}\right) \cdot \left(\frac{2}{n}\right)$$
$$= (-1)^{\frac{m^2-1}{8}} \cdot (-1)^{\frac{n^2-1}{8}} = (-1)^{\frac{m^2-1}{8}+\frac{n^2-1}{8}} = (-1)^{\frac{(mn)^2-1}{8}}$$

For an inductive proof of the last formula, say on the number of prime factors, begin with the fact the quadratic reciprocity holds for prime numbers. Let us suppose that the formula holds with a, b, c, d. Hence the formula (5.2.13) from lemma 45 yields the result:

$$\left(\frac{ab}{cd}\right) = \left(\frac{a}{c}\right) \cdot \left(\frac{b}{d}\right)$$
$$= (-1)^{\frac{a-1}{2}\frac{c-1}{2}} \left(\frac{c}{a}\right) \cdot (-1)^{\frac{b-1}{2}\frac{d-1}{2}} \left(\frac{d}{b}\right)$$
$$= (-1)^{\frac{a-1}{2}\frac{c-1}{2}+\frac{b-1}{2}\frac{d-1}{2}} \left(\frac{cd}{ab}\right)$$
$$= (-1)^{\frac{ab-1}{2}\frac{cd-1}{2}} \left(\frac{cd}{ab}\right)$$

$\square$

5.2.3 Jacobi Symbols in mathematica

In the computer language mathematica, the symbol JacobiSymbol[m, n] is defined for arbitrary integers m and n. Such a symbol is uniquely specified by the following properties

(5.2.20) $$\left(\frac{1}{n}\right) = \left(\frac{m}{1}\right) = 1$$

(5.2.21) $$\left(\frac{0}{n}\right) = 1 \text{ if } n = \pm 1 \qquad \left(\frac{0}{n}\right) = 0 \text{ if } n \neq \pm 1$$

(5.2.22) $$\left(\frac{m}{0}\right) = 1 \text{ if } m = \pm 1 \qquad \left(\frac{m}{0}\right) = 0 \text{ if } m \neq \pm 1$$

(5.2.23) $$\left(\frac{m}{n}\right) = 0 \text{ if } m \text{ and } n \text{ are both even}$$

(5.2.24) $$\left(\frac{m}{n}\right) = \left(\frac{m}{n/4}\right)$$
if $m \neq 0$ and n is divisible by 4

(5.2.25) $$\left(\frac{m}{n}\right) = (-1)^{\frac{m^2-1}{8}} \left(\frac{m}{n/2}\right)$$
if m is odd and n is even

(5.2.26) $$\left(\frac{m}{n}\right) = \left(\frac{m}{-n}\right) \quad \text{if } m > 0 \text{ and } n < 0$$

(5.2.27) $$\left(\frac{m}{n}\right) = -\left(\frac{m}{-n}\right) \quad \text{if } m < 0 \text{ and } n < 0$$

(5.2.28) $$\left(\frac{m}{n}\right) = \left(\frac{m \bmod n}{n}\right) \quad \text{if } n \geq 3 \text{ is odd}$$

(5.2.29) $$\left(\frac{m}{n}\right) = \left(\frac{m/4}{n}\right)$$
if m is divisible by 4 and $n \neq 0$

(5.2.30) $$\left(\frac{m}{n}\right) = (-1)^{\frac{n^2-1}{8}} \left(\frac{m/2}{n}\right)$$
if m is even and n is odd

(5.2.31) $$\left(\frac{m}{n}\right) = (-1)^{\frac{m-1}{2} \frac{n-1}{2}} \left(\frac{n}{m}\right)$$
if $m \geq 1$ is odd and $n \geq 1$ is odd

Remark. The modular reduction (5.2.28) has to be <u>restricted</u>, to bottom odd integer $n \geq 3$. As already indicated in remark 5.2, otherwise the case $n = 1$ would lead to a contradiction. Because of property (5.2.25), all even n would lead to contradictions, too

Here is a recursive procedure for its computation, together with its check for $|m| \leq 300, |n| \leq 300$:

5.2. JACOBI SYMBOLS

```
In[1]:= Clear[mj]; mj[a_, b_] := mj[a, b] = Which[
    a == 1, 1,   b == 1, 1,
    EvenQ[a] && EvenQ[b], 0,
    a > 0 && b < 0, mj[a, -b],
    a < 0 && b < 0, -mj[a, -b],
    a == 0 && b == -1, 1,
    a == 0, 0,
    a == -1 && b == 0, 1,
    b == 0, 0,
    Divisible[b, 4], mj[a, b/4],
    EvenQ[b], (-1)^((a^2 - 1)/8) *mj[a, b/2],
    a >= b, mj[Mod[a, b], b],
    a <= -1, mj[Mod[a, b], b],
    Divisible[a, 4], mj[a/4, b],
    EvenQ[a], (-1)^((b^2 - 1)/8) *mj[a/2, b],
    a < b, (-1)^((a - 1) (b - 1)/4)*mj[b, a],
    True, Indeterminate]

In[2]:= ctimes = Times @@ # &;

In[3]:= mille = 300; probe =
 Boole[Table[JacobiSymbol[x, y] == mj[x, y],
   {x,-mille,mille}, {y,-mille,mille}]];
   ctimes[Flatten[probe]]
Out[3]= 1
```

Under the assumptions that $a \geq 1$ and $b \geq 1$ are both odd, one may simplify the procedure for the calculation of the Jacobi symbol to the case traditionally covered in number theory.

```
In[5]:= Clear[mjpodd];
    mjpodd[a_, b_] := mjpodd[a, b] = Which[
    a == 1, 1,   b == 1, 1,
    a == 0, 0,
    a >= b, mjpodd[Mod[a, b], b],
    a <= -1, mjpodd[Mod[a, b], b],
    Divisible[a, 4], mjpodd[a/4, b],
    EvenQ[a], (-1)^((b^2 - 1)/8) *mjpodd[a/2, b],
    a < b, (-1)^((a - 1) (b - 1)/4)*mjpodd[b, a],
    True, Indeterminate]

In[6]:= mille = 600; probe =
 Boole[Table[JacobiSymbol[x, y] == mjpodd[x, y],
```

```
{x,-mille,mille}, {y,1,mille,2}]];
ctimes[Flatten[probe]]
Out[6]= 1
```

Proposition 41. *The Jacobi-Wolfram symbol is an extension of the Jacobi symbol. Beyond its defining properties, the Jacobi-Wolfram symbol has the following properties:*

$$\left(\frac{ab}{n}\right) = \left(\frac{a}{n}\right) \cdot \left(\frac{b}{n}\right) \tag{5.2.32}$$

$$\left(\frac{a}{mn}\right) = \left(\frac{a}{m}\right) \cdot \left(\frac{a}{n}\right) \tag{5.2.33}$$

$$\left(\frac{m}{n}\right) = 0 \quad \text{if } m \text{ and } n \text{ are not relatively prime} \tag{5.2.34}$$

$$\left(\frac{m}{n}\right) = (-1)^{\frac{m-1}{2} \frac{n-1}{2}} \left(\frac{n}{m}\right) \tag{5.2.35}$$

if m, n are odd and not both negative

$$\left(\frac{m}{n}\right) = -(-1)^{\frac{m-1}{2} \frac{n-1}{2}} \left(\frac{n}{m}\right) \tag{5.2.36}$$

if $m, n \leq -1$ are both odd

$$\left(\frac{2}{m}\right) = \left(\frac{m}{2}\right) = (-1)^{\frac{m^2-1}{8}} \quad \text{if } m \text{ is odd} \tag{5.2.37}$$

$$\left(\frac{-1}{m}\right) = (-1)^{\frac{m-1}{2}} \quad \text{if } m \text{ is odd} \tag{5.2.38}$$

I give only the following hint for a more complete proof.

Proof of equation (5.2.32). We check cases $n = 1, n = 0$ directly

With the help of equations (5.2.26) and (5.2.27) the case with $n < 0$ is reduced to $n > 0$.

With the help of equations (5.2.24) and (5.2.25) the case with n even is reduced to n odd.

With the help of equation (5.2.28) we reduce to the case with $0 \leq a < n, 0 \leq b < n$.

We check the cases $a = 0, 1$ or $b = 0, 1$. With the help of equations (5.2.29) and (5.2.30) we reduce to the case $a \geq 1$ and $b \geq 1$ both odd.

I claim it is enough to check equation (5.2.32) under the additional assumptions that $n \geq 3, a \geq 3$ $b \geq 3$ are all odd and $a < n$ and $b < n$. But that restricted case just deals with the Jacobi symbol. □

5.2.4 Jacobi Symbols and Quadratic Residues

Lemma 46. *Fix some odd number $m \geq 3$. For numbers a which are relatively prime to m, there exist $\phi(m)$ remainder classes modulo m. Either one of the following cases (a) or (b) occurs:*

(a) *The Jacobi symbol is $\left(\frac{a}{m}\right) = 1$ for $\phi(m)/2$ choices of a and $\left(\frac{a}{m}\right) = -1$ for the remaining $\phi(m)/2$ choices of a.*

(b) *The Jacobi symbol is $\left(\frac{a}{m}\right) = 1$ for all $\phi(m)$ choices of a.*

Proof. In the following, we assume $m \geq 3$ to be odd. By formula

$$\left(\frac{ab}{m}\right) = \left(\frac{a}{m}\right) \cdot \left(\frac{b}{m}\right)$$

from proposition 40 the mapping $G_m^* \mapsto \{\pm 1\}$ which takes a to the Jacobi symbol $\left(\frac{a}{m}\right)$ is a group homomorphism. Its kernel is a subgroup

(5.2.39) $$H = \{a \in G_m^* : \left(\frac{a}{m}\right) = 1\}$$

The index $[H, G_m^*]$ is either 2 or 1. This yields the cases (a) or (b), respectively. □

Lemma 47. *The alternative (b) in Lemma 46 occurs if and only if m is a perfect square.*

Problem 71. *Prove Lemma 47. Begin with the special case that m is a prime power. Use induction on the number of different primes in m.*

Proof. We assume the Jacobi symbols are $\left(\frac{a}{m}\right) = 1$ for all a relatively prime to m, and show that m is a perfect square. We proceed by induction on the number t of different primes in m.

Begin with the case $t = 1$. Here $m = p^s$ is a prime power. There exists a primitive root r. Thus G_m^* is a cyclic group generated by the powers of r. Which one of cases (a) or (b) occurs is determined by the value of the Jacobi symbol $\left(\frac{r}{m}\right)$. We claim

(5.2.40) $$\left(\frac{r}{m}\right) = \left(\frac{r}{p^s}\right) = \left(\frac{r}{p}\right)^s = \left(\frac{r_1}{p}\right)^s = (-1)^s$$

with $r_1 \equiv r \bmod p$. Now r_1 is a primitive root modulo prime p and hence r_1 is not a quadratic residue modulo p. Any Jacobi symbol is an extension of the Legendre symbol. Now the Euler criterium and Euler's formula (5.1.2) may be used and yield

$$\left(\frac{r_1}{p}\right) = r_1^{\frac{p-1}{2}} = -1$$

confirming formula (5.2.40). We see that case (a) occurs for s odd, whereas case (b) occurs for s even, and hence if and only if $m = p^s$ is a perfect square.

We proceed now by induction on the number of different primes in the denominator. Assume the assertion to be true for denominators with less than t different prime factors. Let mn have t different prime factors and assume $\gcd(m,n) = 1$. We assume that

$$\left(\frac{x}{mn}\right) = 1 \quad \text{for all } x \text{ with } \gcd(x, mn) = 1,$$

and have to check whether mn is a perfect square. For any a with $\gcd(a,m) = 1$ and b with $\gcd(b,n) = 1$ the Chinese remainder Theorem shows existence of x modulo mn such that

$$x \equiv a \pmod{m}$$
$$x \equiv b \pmod{n}$$

We may choose $b = 1$ and conclude

$$1 = \left(\frac{x}{mn}\right) = \left(\frac{a}{m}\right)\left(\frac{1}{n}\right) = \left(\frac{a}{m}\right)$$

$$\left(\frac{a}{m}\right) = 1 \quad \text{for all } a \text{ with } \gcd(a,m) = 1$$

Hence the induction assumption implies that m is a perfect square.

We may choose $a = 1$ and conclude

$$1 = \left(\frac{x}{mn}\right) = \left(\frac{1}{m}\right)\left(\frac{b}{n}\right) = \left(\frac{b}{n}\right)$$

$$\left(\frac{b}{n}\right) = 1 \quad \text{for all } b \text{ with } \gcd(b,n) = 1$$

Hence the induction assumption implies that n is a perfect square. Since m and n are relatively prime, we conclude that there product mn is a perfect square, too, as to be shown.

Conversely, assume that $m = n^2$ is a perfect square. Then the Jacobi symbols are

$$\left(\frac{a}{m}\right) = \left(\frac{a}{n^2}\right) = \left(\frac{a}{n}\right)^2 = 1$$

for all a relatively prime to m. □

5.2. JACOBI SYMBOLS

Let

$$K = \{a \in G_m^* : x^2 \equiv a \pmod{m} \text{ is solvable }\}$$

denote the remainder classes of a modulo m which are relatively prime to m, and for which the quadratic congruence

(5.2.44) $$x^2 \equiv a \pmod{m}$$

is solvable. The mapping $h : x \in G_m^* \mapsto x^2 \in K$ is a group homomorphism. The image of h is, $K \subset G_m^*$. This is hence a subgroup. The kernel of h is

$$\text{kernel } h = \{x \in G_m^* : x^2 \equiv 1 \pmod{m}\} =: Q$$

By the main theorem of group theory the image of h is isomorphic to the quotient of the domain by the kernel:

$$K \simeq G_m^*/\text{kernel } h$$

From here we get the subgroup index

(5.2.41) $$\text{index } [K, G_m^*] = \dim \text{ kernel } h$$

Proposition 42. *Let $m = 2^s$ be a power of 2. For $s \geq 3$, the quadratic residues are a cyclic subgroup with index $[K, G_m^*] = 4$.*

The cases $m = 2, 4$ are exceptional:
$G_2^* = K = \{1\}$ with index $[K, G_2^*] = 1$; whereas
$G_4^* = \{1, 3\} \sim \mathbf{Z}_2$ with $K = \{1\}$ and index $[K, G_4^*] = 2$

A simple proof. The kernel dim kernel h consists of the four elements $\{1, -1, 1 + \frac{m}{2}, -1 + \frac{m}{2}\}$. I may leave to the reader that these four elements are in the kernel. Conversely, let $x = 1 + a$ and solve $x^2 \equiv 1 \pmod{m}$. One gets $a(2 + a) \equiv 0 \pmod{m}$ which has four solutions:

$$a = 0, x = 1,$$
$$a = -2, x = -1,$$
$$a = \tfrac{m}{2}, x = 1 + \tfrac{m}{2},$$
$$a = -2 + \tfrac{m}{2}, x = -1 + \tfrac{m}{2}$$

Indeed, either one factor is divisible by m, or $a \equiv 2 \pmod{4}, 2+a = m/2$ or $2+a \equiv 2 \pmod{4}, a = m/2$. All four solution are different since $m \geq 8$. By equation (5.2.41) index $[K, G_m^*] = \dim$ kernel $h = 4$. □

Lemma 48 (The kernel of the squaring homomorphism).
Fix some integer $m \geq 2$. Let t denote the number of its different prime factors. The kernel of the squaring homomorphism $h : x \in G_m^* \mapsto x^2$ has the dimension

(i) dim kernel $h = 2^t$ if m is either odd or $m \equiv 4 \pmod 8$;

(ii) dim kernel $h = 2^{t-1}$ if m is even and $m \equiv 2 \pmod 4$;

(iii) dim kernel $h = 2^{t+1}$ if m is even and $m \equiv 0 \pmod 8$.

Proof. Let
$$m = \prod_{1 \leq i \leq t} p_i^{r_i}$$
be the prime factorization of m. The kernel of the squaring homomorphism h can be determined via the Chinese remainder theorem. Indeed any solution of $x^2 \equiv 1 \pmod m$ is obtained by solving at first

(5.2.42) $$x_i^2 \equiv 1 \pmod{p_i^{r_i}}$$

for all $1 \leq i \leq t$ separately, then solving

(5.2.43) $$x \equiv x_i \pmod{p_i^{r_i}} \quad \text{for } 1 \leq i \leq t$$

which is a standard Chinese remainder problem. The equation (5.2.42) has two solutions $x_i = \pm 1$ if either p_i is odd or $p_i^{r_i} = 4$. By lemma 42 the equation (5.2.42) has four solutions if $8 \mid p_i^{r_i}$. But there exists only one solution if $p_i^{r_i} = 2$. Because of equation (5.2.43) these solutions may be combined arbitrarily. Hence the dimension of the kernel is obtained by taking the product of all $1 \leq i \leq t$ and comes out as claimed. □

Proposition 43 (Quadratic residues for composite numbers). *Fix some integer $m \geq 2$. Let t denote the number of its different prime factors. The set K of quadratic residues is a subgroup of the Euler group G_m^*. Its index is*

(i) index $[K, G_m^*] = 2^t$ if m is either odd or $m \equiv 4 \pmod 8$;

(ii) index $[K, G_m^*] = 2^{t-1}$ if m is even and $m \equiv 2 \pmod 4$;

(iii) index $[K, G_m^*] = 2^{t+1}$ if m is even and $m \equiv 0 \pmod 8$.

Especially $K = G_m^*$ if and only if $m = 2$. In all other cases, quadratic non-residues exist.

5.2. JACOBI SYMBOLS

Proof. Because of equation

(5.2.41) $\qquad$ index $[K, G_m^*] = \dim \text{kernel } h$

it is enough to determine the dimension of the kernel, as already obtained by lemma 48. $\qquad\square$

A second direct proof. The proof of proposition 43 is done by induction on the number t of different primes in m.

Begin with the case $t = 1$. Here $m = p^s$ is a prime power. Assume that p is an odd prime. There exists a primitive root r. Thus G_m^* is a cyclic group generated by the powers of r. The quadratic residues are the even powers of r. Hence they are a subgroup $K \subset G_m^*$ of index 2. The special case $p = 2$ is covered with proposition 42. Here one gets a subgroup $K \subset G_{2^s}^*$ of index 1, 2 or 4 in the cases $s = 1, s = 2$ and $s \geq 3$.

We proceed now by induction on the number of different primes. Assume the assertion that the quadratic residues are a subgroup of the Euler group to be true for less than t different prime factors. Let mn have t different prime factors but both m and n less than t prime factors. Too, we assume $\gcd(m, n) = 1$.

To check that $K_{mn} \subset G_{mn}^*$ is a subgroup, we assume that both $c, d \in G_{mn}^*$ are perfect squares modulo mn. Let $a \equiv c \pmod{m}$, $b \equiv c \pmod{n}$, and similarly $a' \equiv d \pmod{m}$, $b' \equiv d \pmod{n}$. We solve at first the quadratic congruences

$$y^2 \equiv aa' \pmod{m}$$
$$z^2 \equiv bb' \pmod{n}$$

This is possible. Indeed both $a, a' \in G_m^*$ are perfect squares modulo m. Invoking the induction assumption, we conclude that their product aa' is a perfect square modulo m, too. Both $b, b' \in G_n^*$ are perfect squares modulo n. Invoking the induction assumption, we conclude that their product bb' is a perfect square modulo n, too. Next we use the Chinese remainder Theorem to get x such that

$$x \equiv y \pmod{m}$$
$$x \equiv z \pmod{n}$$

Together we have obtained a solution of

$$x^2 \equiv y^2 \equiv aa' \equiv cd \pmod{m}$$
$$x^2 \equiv z^2 \equiv bb' \equiv cd \pmod{n}$$

and hence $x^2 \equiv cd \pmod{mn}$ as claimed. In other words, the product to two quadratic residues is again a quadratic residue. Too, we get the index of the group K_{mn} from

$$\text{index } [K_{mn}, G_{mn}^*] = \text{index } [K_m, G_m^*] \cdot \text{index } [K_n, G_n^*]$$
$$= 2^{t(m)} 2^{t(n)} = 2^{t(m)+t(n)} = 2^{t(mn)}$$

as claimed. The modification occurring in the special case $p = 2$ is easy. $\square$

We can now give a very short but not constructive proof of theorem 14.

Short proof of theorem 14. Let $L \subseteq G_m^*$ be the set of a for which assumptions (a), (b1) and (b2) from theorem 14 hold. This set is a subgroup. Moreover index $[H, G_m^*] =$ index $[K, G_m^*]$. Hence $H = K$. In other words $a \in H$ if and only if a is a quadratic residue modulo m. $\square$

Proposition 44 (Jacobi symbols confirm that <u>some</u> quadratic congruences are not solvable). *Let $m \geq 3$ be any odd number and a be an integer. We assume that a and m are relatively prime. If the quadratic congruence*

(5.2.44) $$x^2 \equiv a \pmod{m}$$

is solvable then the Jacobi symbol is $\left(\frac{a}{m}\right) = 1$.

Hence $\left(\frac{a}{m}\right) = -1$ confirms that the congruence (5.2.44) is not solvable.

Remark. In general there exist many numbers a for which the quadratic congruence (5.2.44) is not solvable, but nevertheless the Jacobi symbol turns out to be $\left(\frac{a}{m}\right) = 1$. The converse of proposition 44 is only true for some exceptional choices of m.

Theorem 15. *Let p be an <u>odd prime</u> and $s \geq 1$ be any <u>odd</u> integer. For any integers a and $m = p^s$ relatively prime, the quadratic congruence*

$$x^2 \equiv a \pmod{m} \quad \text{is solvable iff} \quad \left(\frac{a}{m}\right) = 1$$
$$x^2 \equiv a \pmod{m} \quad \text{is not solvable iff} \quad \left(\frac{a}{m}\right) = -1$$

For any integers a and $m = p^s$ the Jacobi symbol is

(5.2.45) $$\left(\frac{a}{m}\right) \equiv a^{\frac{p-1}{2}} \pmod{p}$$

Proposition 45. *If and only if $m = p^s$ is an odd power of an odd prime do the Jacobi symbols yield a necessary and <u>sufficient</u> criterium for solvability of quadratic congruences, as given in Theorem 15.*

Remark. The rather interesting case of even m is to be excluded since the Jacobi symbols may only be defined for odd bottom integer.

Proof. Jacobi symbols yield a necessary and <u>sufficient</u> criterium for solvability of quadratic congruences if and only if the two subgroups

$$K = \{a \in G_m^* : x^2 \equiv a \pmod{m} \text{ is solvable }\}$$

and

$$H = \{a \in G_m^* : \left(\frac{a}{m}\right) = 1\}$$

turn out to be equal. In general, proposition 44 yields only the inclusion $K \subseteq H$. The equality $K = H$ occurs if and only if index $[K, G_m^*] =$ index $[H, G_m^*]$. As shown in proposition 43, the former index $[K, G_m^*] = 2^t \geq 2$ with t denoting the number of prime factors of m. The latter index $[H, G_m^*]$ is either 2 or 1, as shown by lemma 46.

Thus the equality occurs if and only if

$$\text{index } [K, G_m^*] = \text{index } [H, G_m^*] = 2$$

By proposition 43, we see that m is a prime power.
By lemma 47, index $[H, G_m^*] = 2$ implies that m is not a perfect square. Hence we are left with the only possibility that m is an odd power of an odd prime. □

5.3 Primitive Roots

Definition 30 (Primitive root). Let $m > 1$. The integer a is called a *primitive root modulo m* iff the *order of the integer a modulo m* is equal to $\phi(m)$.

In other words, the integer a is primitive root modulo m if and only if

$$a, a^2, a^3, \ldots a^{\phi(m)}$$

are a system of representatives of the congruence classes modulo m that a relatively prime to m.

Proposition 46. *If there exists a primitive root modulo m, the number of such primitive roots in the interval $1 \leq a < m$ is $\phi(\phi(m))$.*

Proof. Let the integer a be a primitive root modulo m. Hence

$$a, a^2, a^3, \ldots a^{\phi(m)}$$

are a system of representatives of the congruence classes modulo m that a relatively prime to m. It is not hard to check that among them, all powers

$$a^j \text{ with } 1 \leq j < m \text{ and } \gcd(j, \phi(m)) = 1$$

are primitive roots, whereas the remaining powers a^j have orders

$$\frac{\phi(m)}{\gcd(j, \phi(m))} < \phi(m)$$

which are lower, and hence are not primitive roots. We have obtained $\phi(\phi(m))$ primitive roots, and they all have different residues modulo m. Hence there are $\phi(\phi(m))$ different primitive roots $b \equiv a^j \pmod{m}$ in the interval $1 \leq b < m$. □

Theorem 16. *Let $m > 1$. There exists a primitive root modulo m in the following cases:*

- *m is $2, 4$ or p^r or $2p^r$ where p is an odd prime and $r \geq 1$.*

There exists no primitive root modulo m in the following cases:

- *m is divisible by 8;*
- *m is divisible by 4 and an odd prime;*
- *m is divisible by two different odd primes.*

5.3.1 Primitive Roots, Mainly for Primes

Gauss has proved that for any prime number there exists a primitive root. See theorem 17 below. The complete proof of Theorem 16 and related matters for prime powers is postponed to section 5.3.2.

The existence part is dealt with in theorem 18 below. The nonexistence part is dealt with in theorem 19 below.

Problem 72. *Assume that no primitive root exists modulo $m > 1$. Show that the Euler totient function $\phi(m)$ is divisible by 4. Is the converse true?*

5.3. PRIMITIVE ROOTS

Answer. A primitive root exists modulo the integer m in the cases where m is $2, 4$ or p^r or $2p^r$ where p is an odd prime and $r \geq 1$. The opposite case that no primitive exists occurs if the integer m is either divisible by 8, or by two different odd primes $p \neq q$. In these cases, it is easy to check that $\phi(m)$ is divisible by 4.

The converse is not true. Let p be any prime with $p \equiv 1 \pmod{4}$ and put $m = p^r$ or $2p^r$. In this case, a primitive root exists and $\phi(m) = p^{r-1}(p-1)$ is divisible by 4, nevertheless.

Proposition 47 (Gauss). *For any $m \geq 1$, the sum of the values $\phi(d)$ over all divisors $d \mid m$ equals m:*

(5.3.1) $$\sum \{\phi(d) : d \text{ divides } m\} = m$$

Problem 73. *Check the formula (5.3.1) for a prime power $m = p^r$.*

Answer.

$$\sum \{\phi(d) : d \mid p^r\} = \sum_{t=0}^{r} \phi(p^t) = 1 + \sum_{t=1}^{r} p^{t-1}(p-1)$$
$$= 1 + \frac{p^r - 1}{p - 1} \cdot (p-1) = p^r$$

Lemma 49. *Assume the integers a and b are relatively prime, and assume that the formula (5.3.1) holds for a and b. Then the formula (5.3.1) holds for the product ab, too.*

Proof. Since $\gcd(a, b) = 1$, any divisor $d \mid ab$ of the product can be factored

$$d = \gcd(ab, d) = \gcd(a, d) \cdot \gcd(b, d) =: e \cdot f$$

Conversely, the product ef of any divisors $e \mid a$ and $f \mid b$ is a divisor of ab. From the multiplicativity of the ϕ-function one gets

$$\sum \{\phi(d) : d \mid ab\} = \sum \{\phi(ef) : e \mid a \text{ and } f \mid b\}$$
$$= \sum \{\phi(e)\phi(f) : e \mid a \text{ and } f \mid b\}$$
$$= \sum \{\phi(e) : e \mid a\} \cdot \sum \{\phi(f) : f \mid b\} = a \cdot b$$

as to be shown. □

End of the proof of Proposition 47. The reasoning uses induction on the number s of different prime factors of $m = p_1^{r_1} \cdot p_2^{r_2} \cdots p_s^{r_s}$. Problem 73 gives the induction start for $s = 1$, and Lemma 49 provides the induction step. □

Gauss' independent proof of Proposition 47. We partition the integers $1, 2, , \ldots, m$ into disjoint classes

$$C_d = \{j : 1 \leq j \leq m \text{ and } \gcd(j, m) = d\}$$

for all divisors $d \mid m$. Since $\gcd(j, m) = d$ if and only if $\gcd(j/d, m/d) = 1$, the possible values for j/d are just the integers among $1, 2, \ldots m/d$ which are relatively prime to m/d. Hence $|C_d| = \phi(m/d)$ and counting yields

$$m = \sum \{|C_d| : d \text{ divides } m\} = \sum \{\phi(\frac{m}{d}) : d \text{ divides } m\}$$
$$= \sum \{\phi(d) : d \text{ divides } m\}$$

□

Problem 74. *As an example, put $m = 12$ and determine the classes C_d.*

Answer.

divisor d	C_d	$\phi\left(\frac{12}{d}\right)$
1	$\{1, 5, 7, 11\}$	$\phi(12) = 4$
2	$\{2, 10\}$	$\phi(6) = 2$
3	$\{3, 9\}$	$\phi(4) = 2$
4	$\{4, 8\}$	$\phi(3) = 2$
6	$\{6\}$	$\phi(2) = 1$
12	$\{12\}$	$\phi(1) = 1$

Theorem 17 (Gauss). *For any prime number there exists a primitive root.*

Proof. Let p be an odd prime number, and let the integer a represent any of the residue classes modulo p among the numbers $1, 2, \ldots, p-1$.

The Little Fermat Theorem tells that $a^{p-1} \equiv 1 \pmod{p}$. But $p-1$ need not be the smallest exponent with that property. According to Definition 19, the order of a is the smallest positive integer ω such that

(5.3.2) $$a^\omega \equiv 1 \pmod{p}$$

All divisors $\omega \mid p - 1$ can occur as the order of the integer a.

5.3. PRIMITIVE ROOTS

Let $c(\omega)$ count, among the numbers $1, 2, \ldots, p-1$, the residue classes modulo p having the order ω. Counting all different cases yields

(5.3.3) $$p - 1 = \sum \{\, c(\omega) \;:\; \omega \mid p-1 \,\}$$

Next we show the inequality

(5.3.4) $$c(\omega) \leq \phi(\omega)$$

In the case that $c(\omega) = 0$, we are obviously ready. Assume that $c(\omega) > 0$ and let h be a residue class of order ω. It is not hard to check that all powers

(5.3.5) $$h^j \text{ with } 1 \leq j < \omega \text{ and } \gcd(j, \omega) = 1$$

are $\phi(\omega)$ different residue class of order ω. Can there be any more residue classes of the same order ω? From Lagrange's Theorem 11, we conclude that the equation (5.3.2) has at most ω roots. It is easy to see that these roots are $h, h^2, \ldots h^\omega$. Hence we have obtained indeed all roots. But among them only the roots given by formula (5.3.5) actually have the order ω, the order of the remaining roots is a proper divisor of ω. Hence we conclude that there are exactly $\phi(\omega)$ different residue class of order ω, confirming the estimate (5.3.4).

The "coup de grâce" is now relating the counting formula (5.3.3) and estimate (5.3.4) with equation (5.3.1) from Gauss' Proposition. Finally one obtains

$$p - 1 = \sum \{\, c(\omega) \;:\; \omega \mid p-1 \,\} \leq \sum \{\, \phi(\omega) \;:\; \omega \mid p-1 \,\} = p - 1$$

These chain of inequalities can only be true with equality everywhere. Hence we conclude that $c(\omega) = \phi(\omega)$ for all divisors $\omega \mid p-1$. We get the Corollary 24 below and especially the existence of a primitive root. □

Corollary 24. *Let p be any odd prime number.*

- *There exists residue classes for all orders $\omega \mid p-1$ dividing $p-1$;*

- *for each divisor there exist exactly $\phi(\omega)$ residue classes of order ω;*

- *especially there exist exactly $\phi(p-1)$ primitive roots modulo the prime number p.*

5.3.2 Primitive Roots for Composite Numbers

Proposition 48. *Let p be an odd prime number. Equivalent are*

(i) *g is a primitive root modulo p and $g^{p-1} \not\equiv 1 \pmod{p^2}$;*

(ii) *g is a primitive root modulo p^2;*

(iii) *for every $r \geq 2$, the number g is a primitive root modulo p^r.*

Theorem 18. *Assume that g is a primitive root modulo the odd prime p. There exists an integer t such that $g + tp$ is a primitive root modulo $2p^r$ for all $r \geq 1$.*

We shall prove the more general proposition 49 about the order of an integer to which proposition 48 is an immediate corollary. In the following, I denote by a any integer, and by g a primitive root.

Proposition 49. *Let p be an odd prime number, not dividing the number a.*
Equivalent statements are

(i) *a has the order ω modulo p and $a^\omega \not\equiv 1 \pmod{p^2}$;*

(ii) *a has the order $p\omega$ modulo p^2;*

(iii) *for every $r \geq 2$, the number a has the order $p^{r-1}\omega$ modulo p^r.*

Lemma 50 (Main lemma about orders). *Let p be an odd prime, assume the integer a is not divisible by p and ω is the order of integer a modulo p. Then there exists an integer $a+tp$ which has the order $p\omega$ modulo p^2. Indeed one obtains $p-1$ numbers $a+tp$ with $0 \leq t < p, t \neq t_0$ for which $a+tp$ has the order $p\omega$ modulo p^2.*

Proof. For any choice of the integer t, we get $(a+tp)^\omega \equiv 1 \pmod{p}$. Lemma 54 yields $(a+tp)^{p\omega} \equiv 1 \pmod{p^2}$. We conclude that the order of $a+tp$ modulo p^2 is either ω or $p\omega$, depending on whether

(i) $(a+tp)^\omega \equiv 1 \pmod{p^2}$ or

(ii) $(a+tp)^\omega \not\equiv 1 \pmod{p^2}$.

Towards getting proposition 49, the goal is to obtain the latter alternative (ii). For finding the appropriate value of the variable t,

5.3. PRIMITIVE ROOTS

we use the equivalent congruences to be excluded

$$(a+tp)^\omega \equiv 1 \pmod{p^2}$$
$$a^\omega + \omega a^{\omega-1}tp \equiv 1 \pmod{p^2}$$
$$\frac{a^\omega - 1}{p} \equiv -\omega a^{\omega-1}t \pmod{p}$$
$$a\frac{p-1}{\omega}\frac{a^\omega - 1}{p} \equiv -(p-1)a^\omega t \pmod{p}$$
$$t_0 := a\frac{p-1}{\omega}\frac{a^\omega - 1}{p} \equiv t \pmod{p}$$

The assumptions conspire in a way that the left-hand side t_0 of the last equivalence is indeed an integer. Finally, we choose $t \not\equiv t_0$ (mod p) and arrive at $(a+tp)^\omega \not\equiv 1 \pmod{p^2}$, which in turn implies that the order of $a+tp$ modulo p^2 is $p\omega$ as claimed. □

Proposition 50. *Let p be an odd prime and assume the integer g is a primitive root modulo p. Then there exists indeed $p-1$ numbers $g+tp$ for which $g+tp$ is a primitive root modulo p^2. These good cases are given by $0 \le t < p$ and $t \not\equiv t_0$ from equation (5.3.6). Moreover, there exists a number t such that $g+tp$ is a primitive root modulo $2p^2$.*

Lemma 51. *Let p be an odd prime and let $1 < e \,|\, \omega$ be any divisor of the order ω. Under these assumptions, $p \nmid a^{\omega/e} - 1$ implies $p \nmid a^{p^s \omega/e} - 1$ for all $s \ge 0$.*

Proof. We argue by contradiction: Assume $p \mid a^{p^s \omega/e} - 1$. Now Little Fermat's $p \mid a^{p-1} - 1$ and the little proposition 26 imply $p \mid a^{\gcd(p^s \omega/e, p-1)} - 1 = a^{\omega/e} - 1$ contradicting the assumption $p \nmid a^{\omega/e} - 1$ □

End of the proof of Proposition 49. A merry go around (i) implies item (iii) implies item (ii) implies item (i). Item

(i) a has the order ω modulo p and $a^\omega \not\equiv 1 \pmod{p^2}$;

is equivalent to

(i') $p \,\|\, a^\omega - 1$ and $p \nmid a^{\omega/e} - 1$ for all divisors $1 < e \,|\, \omega$.

By means of Lemma 53 and Lemma 51, we inductively conclude

(is) $p^r \,\|\, a^{p^{r-1}\omega} - 1$, and $p \nmid a^{p^{r-1}\omega/e} - 1$ for all divisors e with $1 < e \,|\, \omega$, hold for all $r \ge 1$.

Clearly this implies

(iii) for every $r \geq 2$, the number a has the order $p^{r-1}\omega$ modulo p^r.

Especially with $r = 2$ we get

(ii) a has the order $p\omega$ modulo p^2;

Hence item (iii) implies item (ii). Finally item (ii) implies item (i). Indeed $p^2 \mid a^{p\omega} - 1$ and Little Fermat's $p \mid a^{p-1} - 1$ by the little proposition imply $p \mid a^{\gcd(p\omega, p-1)} - 1 = a^\omega - 1$. Too, $a^\omega \not\equiv 1 \pmod{p^2}$ holds since a has the order $p\omega$, but not ω modulo p^2. □

End of the proof of Proposition 48. We are now investigating primitive roots is the most interesting case. We use Proposition 49 is the special case $\omega = p - 1$ and $g := a$ In this case

$$(5.3.6) \qquad t_0 := \frac{g^p - ga}{p} \pmod{p}$$

The interesting alternative $t \not\equiv t_0 \pmod{p}$ implies $g + tp$ to be a primitive root modulo p^2, and hence modulo all prime powers p^r, by Proposition 49. There are $p - 1$ choices for t leading to this alternative.

To determine a primitive root modulo double prime powers $2p^r$, one needs an odd primitive root. In at least one among these choices, the number $g + tp$ is odd, and hence we have even obtained a primitive root modulo $2p^r$. An effective procedure is explained in the remark 5.3.3 below. □

End of the proof of Theorem 18. The existence of a primitive root was shown for a prime in Gauss' Theorem 17; for an odd prime power in Proposition 48.

We have covered the cases of the double of an odd prime power by remark 5.3.3. Extra care is needed in the case $m = 2 \cdot 3^r$, and especially $m = 18$. We have seen in Problem 75 that 5 is a primitive root modulo $2 \cdot 3^r$ for all $r \geq 2$. □

Theorem 19. *There exists no primitive root modulo m in the following cases:*

- *m is divisible by 8;*

- *m is divisible by 4 and an odd prime;*

5.3. PRIMITIVE ROOTS

- m is divisible by two different odd primes.

Lemma 52. *Equivalent are*

- *the group $\mathcal{U}(\mathbf{Z}_m)$ is cyclic;*
- *there exists a primitive root modulo m;*
- $\phi(m) = \lambda(m)$;

Proof. Invoking lemma 52, the nonexistence of a primitive root in the remaining cases after having checked that $\lambda(m) \leq \phi(m)/2$, This is done as follows.

- If $m = 2^r$ is divisible by 8, Lemma 58 confirms that the group $\mathcal{U}(\mathbf{Z}_n)$ consists of the two cycles of order 2^{r-2} and hence is not cyclic. No primitive root exists. Too, $\lambda(m) = m/4 = \phi(m)/2$.

- If m is divisible by 4 and an odd prime; let $m = 4 \cdot q$ with $q \geq 3$ odd.

$$\lambda(m) = \mathrm{lcm}[\lambda(4), \lambda(q)] = \mathrm{lcm}[2, \phi(q)] = \phi(q) = \phi(m)/2$$

- If m is divisible by 8 and an odd prime; let $m = 2^r \cdot q$ with $r \geq 3$ and $q \geq 3$ odd.

$$\lambda(m) = \mathrm{lcm}[\lambda(2^r), \lambda(q)] = \mathrm{lcm}[2^{r-2}, \phi(q)]$$
$$\leq 2^{r-2}\phi(q) = \phi(m)/2$$

- m is divisible by two different odd primes. For simplicity, I just assume $m = 2^r p^s q^t$ we two different odd primes p and q and leave the general case to the reader.

$$\lambda(m) = \mathrm{lcm}[\lambda(2^r), \lambda(p^r), \lambda(q^t)]$$
$$= \mathrm{lcm}[\lambda(2^r), (p-1)p^{r-1}, (q-1)q^{t-1}]$$
$$\leq \lambda(2^r) \frac{(p-1)(q-1)}{2} p^{r-1} q^{t-1} \leq \phi(m)/2$$

□

We proceed to the complete proofs. Given any prime power p^r and integer a, it is convenient to introduce the notation

$$p^r \| a :\Leftrightarrow p^r \mid a \text{ but } p^{r+1} \nmid a$$

Lemma 53. *Assume p is an odd prime, the natural number $c \neq 1$ and $p \mid c - 1$. Then*
$$p \, \| \, \frac{c^p - 1}{c - 1}$$

Proof. By assumption $c = 1 + kp$. We use the geometric series and the binomial theorem to calculate modulo p^2:

$$\frac{c^p - 1}{c - 1} = \sum_{j=0}^{p-1} c^j = \sum_{j=0}^{p-1} (1 + kp)^j \equiv \sum_{j=0}^{p-1} (1 + jkp)$$

$$= p + kp \sum_{j=0}^{p-1} j = p + kp \cdot \frac{p(p-1)}{2} \equiv p \pmod{p^2}$$

Hence the quotient is divisible by p, but not by p^2. □

Lemma 54. *Let either p be an odd prime and $n \geq 1$. Under these assumptions, $p^n \, \| \, a^\omega - 1$ implies $p^{n+s} \, \| \, a^{p^s \omega} - 1$ for all $s \geq 0$.*

Proof. The procedure is similar to problem 41. We use induction by $s \geq 0$. The induction start $s = 0$ has been assumed to hold. Here is the induction step $s - 1 \to s$:
We put $c - 1 := a^{p^{s-1}\omega} - 1$ which by induction assumption is maximally divisible by p^{n+s-1}. We have to check maximal divisibility by p^{n+s} for the expression

$$a^{p^s \omega} - 1 = c^p - 1 = \frac{c^p - 1}{c - 1} \cdot (c - 1)$$

By lemma 53 the first factor is maximally divisible by p. By induction assumption the second factor is maximally divisible by p^{n+s-1}. Hence
$$p^{n+s} \, \| \, a^{p^s \omega} - 1$$
as to be shown. □

5.3.3 Primitive Roots for some Examples

Remark. We assume that a (small) primitive root a modulo prime p is known, for example from a table. To determine a (small) primitive root modulo prime powers p^r, we need just to distinguish two cases:

(i) $p^2 \nmid a^p - a$. This is indeed the most common case. We see that $t_0 \not\equiv 0 \pmod{p}$. Hence we can put $t = 0$ and conclude that integer a is a primitive root modulo all prime powers p^r.

5.3. PRIMITIVE ROOTS

(ii) $p^2 \mid a^p - a$. This is indeed a rare exception. We see that $t_0 \equiv 0 \pmod{p}$. Hence we can put $t = 1$ and conclude that $a + p$ is a primitive root modulo all prime powers p^r.

Remark. To determine a (small) primitive root modulo double prime powers $2p^r$, one needs at first an <u>odd</u> primitive root. This can be obtained as follows:

(i) $p^2 \nmid a^p - a$ **and a is odd.** We see that $t_0 \not\equiv 0 \pmod{p}$. Hence we can put $t = 0$ and conclude that a is a primitive root modulo all double prime powers $2p^r$;

(ii) $p^2 \mid a^p - a$ **and a is odd.** We see that $t_0 \equiv 0 \pmod{p}$. Hence we can put $t = 2$ and conclude that $a+2p$ is a primitive root modulo double all prime powers $2p^r$.

(iii) $t_0 \not\equiv 1 \pmod{p}$ **and a is even.** Hence we can put $t = 1$ and conclude that $a+p$ is a primitive root modulo all double prime powers $2p^r$;

(iv) $t_0 \equiv 1 \pmod{p}$ **and a is even.** Hence we can put $t = 3$ and conclude that $a + 3p$ is a primitive root modulo all double prime powers $2p^r$;

Problem 75. *We take from a common table that 2 is a primitive root modulo 3. Check whether 2 is a primitive root modulo 9. Find a primitive root modulo $2 \cdot 3^r$ for all $r \geq 2$.*

Problem 76. *We take from a common table that 2 is a primitive root modulo 11. Check whether 2 is a primitive root modulo 121. Is 2 a primitive root modulo prime powers 11^r for $r > 2$?*

Answer. We calculate t_0 from above:

$$\frac{2^{11} - 2}{11} = 186 \equiv 10 \pmod{11}$$

and hence $t_0 = 10$ is nonzero. The order of 2 modulo 121 is <u>not</u> 11, but 121. Hence 2 is a primitive root modulo 121. By Proposition 48 below, we conclude that 2 is a primitive root modulo any prime power 11^r.

Problem 77. *Find a small primitive root modulo $2 \cdot 11^r$ for $r > 2$.*

Answer. Since $a = 2$ and $t_0 = 10$ are both even, we see that $a + p = 13$ is a primitive root modulo all double prime powers $2p^r$.

Remark. Remember the way how Proposition 46 counts the primitive roots: the Proposition gives the number of primitive roots modulo m *in the interval* $1 \leq a < m$.

In the first example, we have the two primitive roots 5 and 11 in the interval $1 \leq a_i < 18$.

For the second example, we get at first four primitive roots $2, 8, 7, 6$ modulo 11. We get the small root 2 from a table. Proposition 46 gives all the primitive roots

$$(2^j \mod 11) \text{ with } j = 1, 3, 7, 9$$

where the exponents j are relatively prime to $\phi(m) = 10$. Explicitly one calculates $2^7 \equiv 7 \pmod{11}$ and $2^9 \equiv 6 \pmod{11}$. In a second step, the primitive roots modulo 121 can be obtained from Proposition 50. The bad exceptional case has to eliminated with equation (5.3.6);— indeed for each of the four primitive roots $2, 6, 7, 8$ modulo 11 <u>separately</u>. For example, in the case $a = 2$ we get

$$t_0 := \left(\frac{a^p - a}{p} \mod 11 \right)$$
$$= \left(\frac{2^{11} - 2}{11} \mod 11 \right) \equiv 186 \equiv 10 \pmod{11}$$

Let $t_0 = \frac{a^{11}-a}{11} \pmod{11}$.

a	t_0	roots mod 121
2	10	$2 + 11t$ with $t = 0 \ldots 9$
8	10	$8 + 11t$ with $t = 0 \ldots 9$
7	3	$7 + 11t$ with $t = 0, 1, 2$ and $t = 4 \ldots 10$
6	8	$6 + 11t$ with $t = 0 \ldots 7$ and $t = 9, 10$

Thus we have obtained the 40 primitive roots in the interval $1 \leq a_i < 121$.

5.3.4 Semiprimitive Roots for Powers of Two

This is my definition:

Definition 31 (Semiprimitive root). Let $r \geq 3$ and $n = 2^r$. I call $a \in \mathcal{U}(\mathbf{Z}_n)$ a *semiprimitive root* modulo n if the order of a modulo n equals half the order of the group $\mathcal{U}(\mathbf{Z}_n)$, hence equals $\frac{n}{4}$.

5.3. PRIMITIVE ROOTS

Lemma 55. *Assume the number $c \neq 1$ and $4 \mid c - 1$. Then*

$$2 \, \Big\| \, \frac{c^2 - 1}{c - 1}$$

Lemma 56. *Under the assumption $n \geq 2$, $2^n \, \| \, a^\omega - 1$ implies $2^{n+s} \, \| \, a^{2^s \omega} - 1$ for all $s \geq 0$.*

Proof. The procedure is similar to problem 41. The relevant prime is $p = 2$. We use induction by $s \geq 0$. The induction start $s = 0$ has been assumed to hold. Here is the induction step $s - 1 \to s$ fr $s \geq 1$:

We put $c - 1 := a^{2^{s-1}\omega} - 1$ which by induction assumption is maximally divisible by 2^{n+s-1} and hence divisible by 4. We have to check maximal divisibility by 2^{n+s} for the expression

$$a^{2^s \omega} - 1 = c^2 - 1 = (c+1) \cdot (c-1)$$

Since $c \equiv 1 \mod 4$, the first factor is maximally divisible by 2. By induction assumption the second factor is maximally divisible by 2^{n+s-1} Hence

$$2^{n+s} \, \| \, a^{2^s \omega} - 1$$

as to be shown. □

Proposition 51 (About powers of two). *Let $c \neq 1$ be an integer. Equivalent statements are*

(i) $2^r \, \| \, c - 1$ holds with $r \geq 2$;

(ii) there exists $r \geq 2$, such that $2^s \, \| \, c^{2^{s-r}} - 1$ for all $s \geq r$;

(iii) $4 \mid c-1$ and there exists $r \geq 2$ such that for all $s > r$, the order of c modulo 2^s is 2^{s-r} ;

(iv) $4 \mid c-1$ and there exist $r \geq 2$ and $t > r$ for which the order of c modulo 2^t is 2^{t-r} ;

(v) $4 \mid c-1$ and there exist $r \geq 2$ and $t > r$ such that
$2^{t-1} \, \| \, c^{2^{t-r-1}} - 1$;

(vi) $4 \mid c-1$ and there exist $r \geq 2$ and $t > r$ such that $2^s \, \| \, c^{2^{s-r}} - 1$ for all $r \leq s < t$.

The equivalence of (i) ⇔ (ii) has been shown in lemma 56. One sees immediately that (ii) → (iii) → (iv).

(iv) → (v). Put $c' := c^{2^{t-r-1}}$. The assumptions imply

$$4|c-1|c'-1, \text{ hence } 2\|c'+1;$$
$$2^t|c'^2-1 \text{ and } 2\|c'+1 \text{ imply } 2^{t-1}|c'-1;$$
$$2^t \nmid c'-1 \text{ and } 2^{t-1}|c'-1 \text{ imply } 2^{t-1}\|c'-1 = c^{2^{t-r-1}} - 1.$$

$\square$

(v) → (vi) and (ii). For any $s > r$,

$$4|c-1 \text{ and } 2^s\|c^{2^{s-r}} - 1 \text{ imply } 2^{s-1}\|c^{2^{s-r-1}} - 1.$$

Hence we obtain the assertion backtracking for all $s = t-1, t-2, \ldots, r+1$. $\square$

Lemma 57. *Let $n = 2^r$ and either $a = 5, r \geq 3$ or $a = 3, r \geq 4$.*

$$a^{n/8} \equiv 1 + \frac{n}{2} \pmod{n} \text{ and } a^{n/4} \equiv 1 \pmod{n}$$

Hence the order of a modulo n is $n/4$.

Proof. I put $a = 5$ and use induction on $r \geq 3$. We start with $r = 3$, hence $n = 8$, where the assertion is easily confirmed. Here is the induction step $r \to r+1$, respectively $n \to 2n$:

By the induction assumption $5^{n/8} = 1 + \frac{n}{2} + kn$ with some integer k. Taking the square we get

$$5^{n/4} = (1 + kn + \tfrac{n}{2})^2 = (1+kn)^2 + 2(1+kn)\tfrac{n}{2} + \tfrac{n^2}{4}$$
$$= 1 + 2kn + k^2n^2 + (1+kn)n + \tfrac{n^2}{4}$$
$$= 1 + n + \left[k + n\tfrac{k^2+k}{2} + \tfrac{n}{8}\right]2n \equiv 1+n \pmod{2n}$$

since the last square bracket is an integer. Thus we have confirmed the assertion for n replaced by $2n$, as to be checked. The second congruence follows easily. $\square$

Lemma 58. *For $r \geq 3$ and $n = 2^r$, the group $\mathcal{U}(\mathbf{Z}_n)$ consists of the two cycles, with the representatives modulo n*

$$a, a^2, a^3, \ldots a^{n/4} \text{ and } -a, -a^2, -a^3, \cdots - a^{n/4}$$

Here one can put $a = 5$ or $a = 3$. The two cycles are disjoint and hence they cover the entire group $\mathcal{U}(\mathbf{Z}_n)$.

5.3. PRIMITIVE ROOTS

Remark. The equation $x^2 \equiv 1 \pmod{n}$ has the four solutions ± 1 and $\pm 1 + \frac{n}{2}$. Of these 1 and $1 + \frac{n}{2}$ appear in the first cycle $5, 5^2, 5^3, \ldots 5^{n/4}$ whereas -1 and $-1 + \frac{n}{2}$ appear in the second cycle $-5, -5^2, -5^3, \cdots - 5^{n/4}$.

Problem 78. Check the four cases with $n = 8, 16$ and $a = 3, 5$. Check that the two cycles are disjoint for all $r \geq 3$.

Problem 79. Let $m = 2^s$ be a power of 2 with $s \geq 3$,
Prove that the quadratic residues are a cyclic subgroup consisting of

$$K = \{3^2, 3^4, \ldots, 3^{m/4} \equiv 1\} \sim \mathbf{Z}_{m/8}$$

Hence index $[K, G_m^*] = 4$.

Proposition 52. For $r \geq 2$ and $n = 2^r$, the group $\mathcal{U}(\mathbf{Z}_n)$ is isomorphic to the direct product of a cyclic group G of order $|G| = n/4$ with a cyclic group H of order either $|H| = 1$ for $n = 2, 4$; or $|H| = 2$ for $n \geq 8$.

For $n \geq 4$, the group G has the representatives $3, 3^2, 3^3, \ldots 3^{n/4}$ $\pmod{n}$. Alternatively for all $n \geq 8$, the group G has the representatives $5, 5^2, 5^3, \ldots 5^{n/4}$ $\pmod{n}$. The cyclic group H with elements $+1, -1 \pmod{n}$ appears only for $n \geq 8$ as the second additional factor.

For all powers of two $n = 2^r$ with $r \geq 1$, the order of the entire group is given by the Euler totient function $|\mathcal{U}(\mathbf{Z}_n)| = \phi(2^r)$. But the maximal order of an <u>element</u>, respectively integer modulo n, is given by Carmicheal function, denoted by $\lambda(n)$ and defined in 1912, which takes the values:

$$\lambda(1) = \lambda(2) = 1, \ \lambda(4) = \lambda(8) = 2,$$
$$\lambda(2^r) = \phi(2^r)/2 = 2^{r-2} \quad \text{for } r \geq 3,$$

Lemma 59. For any $m > 1$ and integer a, relatively prime to m, the order modulo m is a divisor of the Carmichael function $\lambda(m)$.

Proof. Let $m = p_1^{r_1} \cdot p_2^{r_2} \cdots p_s^{r_s}$ be the prime decomposition of the integer m. Take any integer a such that $\gcd(a, m) = 1$. For all $i = 1 \ldots s$, the order of a modulo $p_i^{r_i}$ is a divisor of $d_i \mid \phi(p_i^{r_i})$. Hence the order of a modulo m is a divisor of

$$\text{lcm}[d_i : i = 1 \ldots s] \quad \text{divides} \quad \text{lcm}[\phi(p_i^{r_i}) : i = 1 \ldots s] = \lambda(m)$$

as to be shown. □

Proposition 53. *For any $m > 1$, there exists an integer g for which the order modulo m is equal to the Carmichael function $\lambda(m)$.*

Proof. Let $m = p_1^{r_1} \cdot p_2^{r_2} \cdots p_s^{r_s}$ be the prime decomposition of the integer m. We distinguish the cases

(i) m is odd;

(ii) $p_1 = 2$, hence m is even, and $r_1 = 1$;

(iii) $p_1 = 2$, hence m is even, and $r_1 = 2$;

(iv) $p_1 = 2$, hence m is even, and $r_1 \geq 3$.

We consider cases (i) and (iii) first. For all primes $p_1, p_2, \ldots, p_s$ dividing m, let a_i be a primitive root modulo $p_i^{r_i}$. In case (iii), let $a_1 := 3$. By the Chinese Remainder Theorem, there exists a solution of the system of simultaneous congruences

$$(5.3.7) \qquad g \equiv a_i \pmod{p_i^{r_i}} \quad \text{for } i = 1 \ldots s$$

For all $i = 1 \ldots s$, the order of g modulo $p_i^{r_i}$, equals $\phi(p_i^{r_i})$. A fortiori, for all $i = 1 \ldots s$, the order of g modulo m, is a <u>multiple</u> of $\phi(p_i^{r_i})$. Hence the order of g modulo m is a multiple of $\lambda(m) = \mathrm{lcm}[\phi(p_i^{r_i}) : i = 1 \ldots s]$. By Lemma 59, the order modulo m of any integer is a <u>divisor</u> of the Carmichael function $\lambda(m)$. Hence the order of g modulo m is a equal to the Carmichael function $\lambda(m)$.

We consider case (ii). Let a_1 be a primitive root modulo $2p_1^{r_1}$. For the remaining primes $p_2, \ldots, p_s$ dividing m, let a_i be a primitive root modulo $p_i^{r_i}$. Now we argue as above.

We consider case (iv). Let a_1 be an element of order 2^{r_1-2} modulo 2^{r_1}. By Lemma 58, we know that this is the maximal possible order. By Lemma 57 we know that $a_1 = 3$ or $a_1 = 5$ we be a possible choice. For the remaining primes $p_2, \ldots, p_s$ dividing m, let a_i be a primitive root modulo $p_i^{r_i}$. Now we argue as above. $\square$

Proposition 54. *Let $m > 1$ be given. For all divisors $d \mid \lambda(m)$ of the Carmichael function, there exists an integer b such that d is equal to the order $\mathrm{ord}_m(b)$ of b modulo m.*

Proof of Proposition 54. By Proposition 53, there exists an integer g such that $\mathrm{ord}_m(g) = \lambda(m)$. Given any divisor $d \mid \lambda(m)$, we put

$$b := g^{\frac{\lambda(m)}{d}}$$

and check that $\mathrm{ord}_m(b) = d$ as required. $\square$

5.4. THE PRODUCT OVER THE EULER GROUP

Problem 80. *Let $m = 11^2 \cdot 23$. Find the value $\lambda(m)$, and an integer a which has the order $\lambda(m)$ modulo m.*

Answer.

$$\lambda(11^2 \cdot 23) = \text{lcm}\left[\lambda(11^2), \lambda(23)\right] = \text{lcm}\left[110, 22\right] = 220$$

We take from a common table that 2 is a primitive root modulo 11. We have seen in Problem 76 above that 2 is a primitive root modulo 121. We take from a common table that 5 is a primitive root modulo 23.

To find the integer a of maximal order, we solve the Chinese remainder problem: [5]

(ch) $\quad\quad\quad\quad a \equiv 2 \pmod{121}$
$\quad\quad\quad\quad\quad\quad a \equiv 5 \pmod{23}$

One gets the solution $a = 1454 \pmod{2783}$.

This integer a has indeed the order $\lambda(m)$ modulo $m = 2783$. Since $a \equiv 2 \pmod{121}$, and 2 is a primitive root, the order of the integer a modulo 121 is $\phi(121) = 110$.

Secondly, since $a \equiv 5 \pmod{23}$, and 5 is a primitive root, the order of the integer a modulo 23 is $\phi(23) = 22$.

The order of $a = 1454$ modulo $\text{lcm}[121, 23]$ is a multiple of both orders, thus a multiple of $\text{lcm}[110, 22] = \lambda(m)$.

On the other hand, Lemma 59 tells that the order of any integer a modulo-m is always divisor of the Carmichael function $\lambda(m)$.

Together, we conclude that the order of $a = 1454$ modulo $m = 2783$ turns out to be the maximal possible order $\lambda(m) = 220$.

5.4 The Product over the Euler Group

Recall that by definition 16 the *Euler group* G_m^* consists of the remainder classes of a modulo m which are relatively prime to m. The modular multiplication is the group operation. In the familiar proof 4.2.3 of Euler's theorem 9 the product of all elements of the Euler group comes up, but its value is not obtained. We calculate now its value modulo m. It turns out to be either -1 or 1, depending on existence or nonexistence of a primitive root.

---

[5] See Problem 35 above

Problem 81. *Show that in the cases where a primitive root exists, the product*
$$A := b_1 \cdot b_2 \cdots b_{\phi(m)}$$
from the proof of Euler's Theorem satisfies $A \equiv -1 \pmod{m}$.

Answer. The cases $m = 2, 4$ can be checked directly Let g be a primitive root modulo $m = p^r$ (odd prime power) or $m = 2p^r$. The Euler totient function $\phi(m)$ is even. The product A is

$$A = g^1 \cdot g^2 \cdots g^{\phi(m)} = g^{1+2+\cdots+\phi(m)} = g^{\frac{\phi(m)(\phi(m)+1)}{2}}$$
$$\equiv g^{\frac{\phi(m)}{2}} \equiv: h \pmod{m}$$

But $h^2 \equiv 1 \pmod{m}$ by Euler's theorem and $h \not\equiv 1 \pmod{m}$ since g is a primitive root. Since a primitive root is assumed to exist, the equation $h^2 \equiv 1 \pmod{m}$ has only two solutions. Hence $h \equiv -1 \pmod{m}$ as claimed.

Problem 82. *Show that in the cases where no primitive root exists, the product*
$$A := b_1 \cdot b_2 \cdots b_{\phi(m)}$$
occurring in the proof of Euler's Theorem satisfies $A \equiv 1 \pmod{m}$.

Here is a solution of this problem. The set
$$Q := \{a \in G_m^* : a^2 \equiv 1 \mod m\}$$
is the kernel of the squaring homomorphism $h : x \in G_m^* \mapsto x^2$ and hence a subgroup of G_m^*. Under the involution $a_i \mapsto a_i^{-1}$, defined by taking inverses, the elements of Q are the fixed points and the elements of $G \setminus H$ are paired. Hence

(5.4.1) $$\prod_{a \in G_m^*} a \equiv \prod_{a \in Q} a \pmod{m}$$

Lemma 60. *The mapping $a \mapsto i(a) = m - a$ is a second (different) involution $G_m^* \mapsto G_m^*$ with restriction $Q \mapsto Q$. For $m \geq 3$, the involution i has no fixed points.*

Proof. Indeed we calculate modulo m and obtain from the assumption $i(a) = a \in G_m^*$: $m - a \equiv a$ implies $2a \equiv m$ and hence $a \mid m$. Hence $\gcd(a, m) = 1$ implies $a = 1$ and finally $2 = 2a \equiv m \equiv 0 \pmod{m}$ implies $m = 2$. But this special case has been excluded by assuming $m \geq 3$. $\square$

5.4. THE PRODUCT OVER THE EULER GROUP

Now the group Q can be partitioned into pairs $\{a, i(a)\}$. For each pair holds $a(m-a) \equiv -1 \pmod{m}$ and hence we obtain

$$(5.4.2) \qquad \prod_{a \in Q} a \equiv (-1)^{|Q|/2} \mod m$$

From equations (5.4.1) and (5.4.2) we see

Lemma 61. *The product of the elements of the Euler group is*

$$(5.4.3) \qquad \prod_{a \in G_m^*} a \equiv (-1)^{|Q|/2} \mod m$$

and hence either $+1$ or -1 modulo m.

Proposition 55. *Excluding the case $m = 2$ we see: the alternative*

$$\prod_{a \in G_m^*} a = \prod_{a \in Q} a \equiv -1 \mod m$$

is equivalent to $Q = \{\pm 1\}$ which is in turn equivalent to existence of a primitive root. The alternative

$$\prod_{a \in G_m^*} a = \prod_{a \in Q} a \equiv +1 \mod m$$

occurs if and only if the order $|Q|$ is divisible by 4 which is in turn equivalent to nonexistence of a primitive root.

Proof. It is easy to see that $|Q|$ is divisible by 4 unless $Q = \{\pm 1\}$. Indeed, if there exists $a \in Q \setminus \{\pm 1\}$, the set $\{1, -1, a, -a\}$ is a subgroup of Q of order four, and hence $|Q|$ is divisible by 4.

Too, the order of kernel Q of squaring homomorphism $h : x \in G_m^* \mapsto x^2$ has already be determined by lemma 48. Let t denote the number of its different prime factors of $m \geq 3$. From there we see that $\dim Q = 2$ occurs for the cases

(i) $\dim Q = 2^t = 2$ if m is either an odd prime power or $m = 4$;

(ii) $\dim Q = 2^{t-1} = 2$ if $m \equiv 2 \pmod 4$ and $t = 2$ and hence $m = 2p^r$.

Because of theorem 16 these turn just out to be the cases where a primitive root exists. □

Remark. Proposition 55 is indeed a theorem of Gauss, who has sketched the proof in his Disquisitiones Arithmeticae. Too, a detailed proof can be found in Ore's book [6] p.263-267.

It seems rather peculiar that case (1-) does occur if and only if there exists a primitive root. It would be nice to establish the equivalence of (1-) with existence of a primitive root even more directly. How does (1-) imply the existence of a primitive root?

Chapter 6

Around Fermat's Last Problem

6.1 The Square Root of a Complex Number

Proposition 56. *The branch with $\Re\sqrt{z} \geq 0$ for the square root of any complex number $z = x + iy$ is*

$$(6.1.1) \quad \sqrt{x+iy} = \sqrt{\frac{\sqrt{x^2+y^2}+x}{2}} + i\,\mathrm{sign}\,(y)\sqrt{\frac{\sqrt{x^2+y^2}-x}{2}}$$

The square root $w = \sqrt{z}$ of any complex number $z = x + iy$ has two branches. The second branch is $w_2 = -w$.

Reason. Name the square root in question $\sqrt{x+iy} =: u + iv$. Squaring yields $x + iy = (u + iv)^2$. Separate the real- and imaginary part to get

$$x = u^2 - v^2 \,,\; y = 2uv$$

The absolute value squared is

$$x^2 + y^2 = |x+iy|^2 = |u+iv|^4 = (u^2+v^2)^2$$

Add and substrate

$$u^2 + v^2 = \sqrt{x^2 + y^2}$$
$$u^2 - v^2 = x$$
$$u^2 = \frac{\sqrt{x^2 + y^2} + x}{2}$$
$$v^2 = \frac{\sqrt{x^2 + y^2} - x}{2}$$

Of the last two expression, one takes real square roots. $u \geq 0$ has been assumed. One needs still to determine the sign of v. But $y = 2uv$ and $u > 0$ imply $\operatorname{sign} y = (\operatorname{sign} u)(\operatorname{sign} v) = \operatorname{sign} v$. In the special case $u = 0$, both signs of v give a correct result for the square root. This special case corresponds to $y = 0$ and $x \leq 0$—the negative numbers $z \leq 0$. □

Problem 83. *Calculate the square roots:*

(a) $\sqrt{-3 + 4i}$, $\sqrt{-12 + 16i}$, $\sqrt{-48 + 64i}$
(b) $\sqrt{-8 - 6i}$, $\sqrt{-32 - 24i}$, $\sqrt{-128 - 96i}$
(c) $\sqrt{-4 - 3i}$, $\sqrt{-16 - 12i}$, $\sqrt{-64 - 48}$
(d) $\sqrt{-6 + 8i}$, $\sqrt{-24 + 32i}$, $\sqrt{-96 + 128i}$
(e) $\sqrt{2i}$, $\sqrt{8i}$, $\sqrt{32i}$

A Gaussian integer that is the square of another Gaussian integer is called a **perfect** square. *Which of the numbers under the square roots are perfect squares, which are not?*

6.2 Pythagorean Triples

Definition 32 (Pythagorean triples). Any three integers $a \geq 1$, $b \geq 1$ and $c \geq 1$ such that $a^2 + b^2 = c^2$ are called a *Pythagorean triple*. If $\gcd(a, b, c) = 1$ the Pythagorean triple is called *primitive*.

Problem 84. *Show that for any Gaussian integer $u + iv$, the real and imaginary parts and the absolute value of its square $a + ib = (u + iv)^2$ are a Pythagorean triple and*

$$a = u^2 - v^2, \quad b = 2uv, \quad c = u^2 + v^2$$

6.2. PYTHAGOREAN TRIPLES

Answer. Let $a + ib = (u + iv)^2$. Squaring the absolute values gives

$$c^2 = a^2 + b^2 = |a + ib|^2 = |u + iv|^4 = (u^2 + v^2)^2$$

Hence $c = u^2 + v^2$. Separating real- and imaginary parts of $a + ib = (u + iv)^2$ yields the formulas for a and b.

Theorem 20 (Pythagorean triples). *For any primitive Pythagorean triple a, b, c one number among a and b is odd and the other one is even. Supposing that b is even, one obtains*

(6.2.1) $\qquad a = p^2 - q^2, \; b = 2pq \; \text{and} \; c = p^2 + q^2$

where $p > 1$ and $1 \leq q < p$ are integers such that $\gcd(p, q) = 1$ and $2 \nmid p - q$. Consequently, the even number $4 \mid b$ is divisible by 4.

Problem 85. *Explain why the even number of any primitive pythagorean triple is divisible by 4.*

Problem 86. *Explain why for any primitive pythagorean triple a, b, c either a or b, but never c is divisible by 3.*

Solution. Assume $3 \nmid b$. Hence $3 \nmid p$ and $p^2 \equiv 1 \pmod 3$. Similarly $3 \nmid q$ and $q^2 \equiv 1 \pmod 3$. Hence $a = p^2 - q^2 \equiv 1 - 1 = 0 \pmod 3$. Thus either b or a are divisible by 3. For a primitive triple holds $\gcd(a, b) = \gcd(a, c) = \gcd(b, c) = 1$. Hence $3 \nmid c$. $\square$

Problem 87. *Explain why for any primitive pythagorean triple a, b, c either a, b or c is divisible by 5.*

Solution. Assume $5 \nmid b$. Hence $5 \nmid p$ and either $p^2 \equiv 1 \pmod 5$ or $p^2 \equiv 4 \pmod 5$. Similarly $5 \nmid q$ and hence either $q^2 \equiv 1 \pmod 5$ or $q^2 \equiv 4 \pmod 5$. We get $a = p^2 - q^2 \equiv 0 \pmod 5$ except in the cases $p^2 \equiv 1 \pmod 5, q^2 \equiv 4 \pmod 5$ or $p^2 \equiv 4 \pmod 5, q^2 \equiv 1 \pmod 5$. But in these cases $c = p^2 + q^2 \equiv 1 + 4 \equiv 0 \pmod 5$ and hence $5 \mid c$. $\square$

Problem 88. *Find,—tentatively infinitely many,—primitive pythagorean triple a, b, c for which both a and c are prime numbers.*

Remark (What I have found). Since $a = p^2 - q^2 = (p + q)(p - q)$ is prime we conclude $p = q + 1$. Hence $c - b = (p - q)^2 = 1$. The first three examples are $3, 4, 5$ next $5, 12, 13$, next $11, 60, 61$. From the two former problems for all further examples $60 \mid b$. Hence $c^2 = (1 + b)^2 \equiv 1 \pmod{120}$ and $a^2 = c^2 - b^2 \equiv 1 \pmod{120}$ and

158   CHAPTER 6.   AROUND FERMAT'S LAST PROBLEM

either $a, c \equiv -1 \pmod{30}$, $a, c \equiv 1 \pmod{30}$, $a, c \equiv 11 \pmod{30}$ or $a, c \equiv 19 \pmod{30}$.

$$b = \frac{a^2 - 1}{2}, \; c = \frac{a^2 + 1}{2}$$

The difficult part is to get primes for a and c.

But one may now try a few more examples for a, b, c: Here is a DrRacket program

```
(require math/number-theory)
(define Pythtwo (lambda(a)
   (let* ( [b (/(sub1 (sqr a))2)] [c (add1 b)])
      (if (prime? c) (list a b c) "") )))

(define (Pythlist p0) (build-list 100
   (lambda (x)
    (let ([p (+ p0(* 30 x))])
      (if (prime? p) (Pythtwo p) "")))))

(define (result p0)
     (newline (current-output-port))
     (fprintf (current-output-port)
    "Pythagorean triples ~a mod 30: "p0 )
     (newline (current-output-port))
     (displayln (Pythlist p0) ))

(result 29)(result 31)(result 11) (result 19)
```

Pythagorean triples 29 mod 30:
   ((29 420 421) (59 1740 1741) (449 100800 100801)
(569 161880 161881) (929 431520 431521) (1439 1035360 1035361)
(1499 1123500 1123501) (1709 1460340 1460341)
(1949 1899300 1899301) (2459 3023340 3023341)
(2549 3248700 3248701) (2609 3403440 3403441)
(2729 3723720 3723721) (2789 3889260 3889261)
(2819 3973380 3973381))
   Pythagorean triples 31 mod 30:
   ((61 1860 1861) (181 16380 16381) (271 36720 36721)
(571 163020 163021) (631 199080 199081) (661 218460 218461)
(751 282000 282001) (991 491040 491041)
(1051 552300 552301) (1171 685620 685621)
(1531 1171980 1171981) (1741 1515540 1515541)

6.2. PYTHAGOREAN TRIPLES

(1831 1676280 1676281) (2161 2334960 2334961)
(2281 2601480 2601481) (2341 2740140 2740141)
(2671 3567120 3567121) (2731 3729180 3729181)
(3001 4503000 4503001))

Pythagorean triples 11 mod 30:

((11 60 61) (71 2520 2521) (101 5100 5101) (131 8580 8581)
(461 106260 106261) (521 135720 135721) (641 205440 205441)
(821 337020 337021) (881 388080 388081) (1031 531480 531481)
(1091 595140 595141) (1151 662400 662401)
(1181 697380 697381) (1361 926160 926161)
(1811 1639860 1639861) (1901 1806900 1806901)
(2351 2763600 2763601) (2381 2834580 2834581)
(2591 3356640 3356641) (2711 3674760 3674761))

Pythagorean triples 19 mod 30:

((19 180 181) (79 3120 3121) (139 9660 9661) (199 19800 19801)
(349 60900 60901) (379 71820 71821)
(409 83640 83641) (739 273060 273061)
(1039 539760 539761) (1069 571380 571381)
(1129 637320 637321) (1459 1064340 1064341)
(1489 1108560 1108561) (2239 2506560 2506561)
(2269 2574180 2574181) (2389 2853660 2853661)
(2539 3223260 3223261) (2659 3535140 3535141)
(2719 3696480 3696481))

Corollary 25. *For any primitive Pythagorean triple a, b, c with b even, we obtain $a + ib = (p + iq)^2$. Thus the Gaussian integer $a + ib$ is a perfect square of an Gaussian integer $p + iq$.*

Proof. Assume $a^2 + b^2 = c^2$ is a primitive triple and let $g := \gcd(a, b)$. To confirm that $g = 1$, we assume that for some prime $r \mid g$ holds $r \mid a$ and $r \mid b$. Hence $r \mid c^2$ and finally $r \mid c$. Thus the triple a, b, c could not be primitive.

To see that a and b cannot both be odd, we calculate modulo 8. For the square of any odd number n holds $n^2 \equiv 1 \pmod 8$. Supposing that a and b are both odd, one obtains $c^2 = a^2 + b^2 \equiv 2 \pmod 8$ which is impossible to hold for a perfect square. Since b is even, we conclude that $a = p^2 - q^2$ is odd. Hence $2 \nmid p - q$. Moreover holds $2g \nmid a$ but $4g \mid b$. □

Corollary 26. *For any Pythagorean triple a, b, c one among a and b is divisible by a higher power of 2 than the other one. Supposing that this number is b, we get*

(6.2.2) $\qquad a = g(p^2 - q^2), \ b = 2gpq \ \text{and} \ c = g(p^2 + q^2)$

with $\gcd(p,q) = 1$ and $2 \nmid p - q$.

Proof. Assume $a^2 + b^2 = c^2$ and let $g := \gcd(a,b)$. For any prime $r \mid g$ holds $r \mid a$ and $r \mid b$. Hence $r \mid c^2$ and finally $r \mid c$. We obtain the Pythagorean triple $\frac{a}{r}, \frac{b}{r}, \frac{c}{r}$. After finitely many divisions of this type one arrives at a primitive triplet

$$a' = \frac{a}{g},\ b' = \frac{b}{g},\ c' = \frac{c}{g}$$

for which $\gcd(a',b') = 1$ and hence $\gcd(a,b) = g$. Now the representation from equation (20) may be used. It may be written in the compact form $a' + ib' = (p + iq)^2$ with $\gcd(p,q) = 1$ and $2 \nmid p - q$. Hence $a + ib = g(p + iq)^2$ and equation (6.2.5) are confirmed. $\square$

Proof of theorem 20. Assume $a^2 + b^2 = c^2$ is a primitive triplet. We have already seen that $\gcd(a,b) = 1$ and that among the numbers a and b one is odd, the other one is even. We assume that b is even and use the equation

$$\frac{c+a}{2} \cdot \frac{c-a}{2} = \frac{c^2 - a^2}{4} = \left(\frac{b}{2}\right)^2$$

Lemma 62. *If the product of two relatively prime factors is a perfect square, then both factors are perfect squares.*

Since the factors $\frac{c+a}{2}$ and $\frac{c-a}{2}$ are relatively prime, we conclude that both factors are perfect squares. Thus there exist positive integers p and q such that

(6.2.3) $$p^2 = \frac{c+a}{2} \text{ and } q^2 = \frac{c-a}{2}$$

Adding and subtracting these formulas yields $a = p^2 - q^2$ and $c = p^2 + q^2$ as claimed. Moreover $4p^2q^2 = b^2$ implies $b = 2pq$.

Let $d := \gcd(p.q)$. Hence $d^2 \mid a$, $d^2 \mid b$ and $d^2 \mid c$. Since the triple is assumed to be primitive, we conclude $d = 1$. Too, p and q cannot both be odd. Otherwise one would get $a \equiv 0 \pmod 8$ and $2 \mid b$ contradicting $\gcd(a,b) = 1$. Hence $2 \nmid p - q$ as to be shown. $\square$

Problem 89. *Generalize the Lemma 62 to the product of three or more factors. Which one of the following conjectures is true, which one is false:*

6.2. PYTHAGOREAN TRIPLES

> "If the product of three pairwise relatively prime factors is a perfect square, then all three factors are perfect squares."

> "If the product of three relatively prime factors is a perfect square, then the product of any two of them is a perfect square."

Definition 33 (Gaussian perfect square). A Gaussian integer $a+ib \in \mathbf{Z}+i\mathbf{Z}$ is called a *Gaussian perfect square* if and only if there exist $u, v \in \mathbf{Z}$ such that $a + ib = e(u + iv)^2$ with $e \in \{1, i, -1, -i\}$.

Remark. If $e = -1$ or $e = -i$, we may use $-(u+iv)^2 = (iu-v)^2$ to produce a relation $a + ib = e'(p+iq)^2$ with $e' \in \{1, i\}$.

If $a + ib$ is a 1-perfect square, then the conjugate $a - ib$ is an 1-perfect square, too. If $a+ib$ is a i-perfect square, then the conjugate $a - ib$ is an i-perfect square, too. Indeed $a+ib = i(u+iv)^2$ implies $a - ib = -i(u-iv)^2 = i(v+iu)^2$

If $a+ib$ is a 1-perfect square, then $b+ia$ is an i-perfect square. Indeed $a + ib = (u + iv)^2$ implies $b + ia = i(u-iv)^2$

Problem 90. *Show that a nonzero Gaussian integer cannot be both a 1-perfect square and a i-perfect square.*

Solution. Assume $(a+ib)^2 = i(c+id)^2$ with integer a, b, c, d. The real and imaginary part and the absolute value yield

$$a^2 - b^2 = -2cd, \quad 2ab = c^2 - d^2 \quad \text{and} \quad a^2 + b^2 = c^2 + d^2$$

Hence $(a+b)^2 = 2c^2$ and the uniqueness of prime factorization implies $a = -b$ and $c = 0$. Similarly, one gets $(c-d)^2 = 2a^2$ and the uniqueness of prime factorization implies $c = d$ and $a = 0$. In the end $a = b = c = d = 0$. □

Example 6.2.1. *These are 1-perfect squares*

(a) $\quad \sqrt{-3+4i} = 1+2i, \quad \sqrt{-12+16i} = 2+4i,$
$$\sqrt{-48+64i} = 4+8i, \dots$$

(b) $\quad \sqrt{-8-6i} = -1+3i, \quad \sqrt{-32-24i} = -2+6i,$
$$\sqrt{-128-96i} = -4+12i, \dots$$

On the other hand, those are not 1-perfect squares:

$$\sqrt{-4+3i} = \frac{(1+3i)\sqrt{2}}{2}$$

$$\sqrt{-6+8i} = (1+2i)\sqrt{2}$$

but they are i-perfect squares, nevertheless.
$$-4 + 3i = i(1 + 2i)^2$$
$$-6 + 8i = i(1 + 3i)^2$$
Here is the most simple example:

(a) $\qquad \sqrt{4} = 2, \ \sqrt{16} = 4, \ \sqrt{64} = 8, \ldots$

(b) $\qquad \sqrt{2i} = 1 + i, \ \sqrt{8i} = 2 + 2i, \ \sqrt{32i} = 4 + 4i, \ldots$

Theorem 21 (Gaussian perfect square). *Any nonzero Gaussian integer $a + ib$ which is a Gaussian perfect square satisfies*

(a) *Either $ab = 0$ or $|a|, |b|, c$ with $c = |a + ib|$ are Pythagorean triple.*

(b) *The common divisor $g := \gcd(a, b) > 0$ is either a perfect square or twice a perfect square.*

Conversely, the two assumptions (a) *and* (b) *imply that the nonzero Gaussian integer $a + ib$ with $ab \neq 0$ is a Gaussian perfect square.*

Remark. Under the additional assumption that $\gcd(a,b) \leq 2$ and calculating (mod 4), indeed only the following six possibilities occur for Gaussian perfect squares.
Let $a + ib = e(p + iq)^2$ and $\gcd(a, b) = f$. a, b, c in the table below are given modulo 4.

a	b	c	e	f	example
1	0	1	1	1	$-3 + 4i = (1 + 2i)^2$
3	0	1	1	1	$3 + 4i = (2 + i)^2$
2	0	2	i	2	$6 + 8i = 2(2 + i)^2 = i(3 - i)^2$
0	1	1	i	1	$4 - 3i = i(1 - 2i)^2 = i(-3 - 4i)$
0	3	1	i	1	$4 + 3i = i(2 - i)^2 = i(3 - 4i)$
0	2	2	1	2	$8 + 6i = 2i(2 - i)^2$ $= [(1 + i)(2 - i)]^2 = (3 + i)^2$

The criterium for 1-perfect square is even more complicated,—and omitted.

Proof of necessity in theorem 21. Let $a + ib$ be a nonzero Gaussian integer that is a Gaussian perfect square. We have assumed $ab \neq 0$. By the remark we may restrict the proof to the case that $a + ib = (p + iq)^2$. Put $c = p^2 + q^2$. One see easily that $c^2 = a^2 + b^2$ confirming item (a). Separating real and imaginary parts and taking the absolute value yields

$$a = (p^2 - q^2), \ b = 2pq \ \text{and} \ c = (p^2 + q^2)$$

6.2. PYTHAGOREAN TRIPLES

Lemma 63. *For $g := \gcd(a,b) > 0$ and $d := \gcd(p,q) > 0$ hold either $g = d^2$ or $g = 2d^2$. Hence $g = fd^2$ holds with either $f = 1$ or $f = 2$.*

Proof. Clearly $d^2 \mid g$. Let

$$p' := \frac{p}{d}, \quad q' := \frac{q}{d}, \quad a' := \frac{a}{d^2}, \quad b' := \frac{b}{d^2}, \quad c' := \frac{c}{d^2}$$

and hence

(6.2.4) $\quad a' = f(p'^2 - q'^2), \; b' = 2fp'q' \text{ and } c' = f(p'^2 + q'^2)$

with $\gcd(a', b') = f = \frac{g}{d^2} > 0$ and $\gcd(p', q') = 1$. Suppose toward a contradiction that any odd prime power r divides f. Then

$$r^2 \mid a'^2 + b'^2 = (p'^2 + q'^2)^2 \text{ and hence } r \mid p'^2 + q'^2$$

Too, $r \mid a' = p'^2 - q'^2$. Hence $r \mid 2p'^2$ and $r \mid p'$. Similarly one gets $r \mid 2q'^2$ and $r \mid q'$. Hence $r \mid \gcd(p', q') = 1$ which is impossible. We conclude that f is a power of 2.

Next we suppose towards a contradiction that $4 \mid f$. Then

$$16 \mid a'^2 + b'^2 = (p'^2 + q'^2)^2 \text{ and hence } 4 \mid p'^2 + q'^2$$

Too, $4 \mid p'^2 - q'^2$. Hence $4 \mid 2p'^2$ and $2 \mid p'$. Similarly hold $4 \mid 2q'^2$ and $2 \mid q'$. Hence $2 \mid \gcd(p', q') = 1$ which is impossible. We conclude that $4 \nmid f$ and thus either $f = 1$ or $f = 2$. □

Thus item (b) has been confirmed, too. □

Proof of sufficiency in theorem 21. We now assume items (a) and (b) to hold. By corollary 26 we get that for the Pythagorean triple a, b, c one number among a and b is divisible by a higher power of 2 than the other one. Supposing that this is b, we get

(6.2.5) $\quad a = g(p^2 - q^2), \; b = 2gpq \text{ and } c = g(p^2 + q^2)$

with $\gcd(p, q) = 1$ and $2 \nmid p - q$. Since we may switch a and b we have made only a restriction easily to be removed. By assumption (b) holds $g = fd^2$ with either $f = 1$ or $f = 2$.

Lemma 64. *Take the case that $g = d^2$ and $f = 1$. Then $2 \nmid p - q$. Thus among p and q one number is odd, the other one is even. We obtain in complex notation as claimed*

$$(a + ib) = (dp + idq)^2$$

Lemma 65. *Take the case that $g = 2d^2$ and $f = 2$. Then $2 \nmid p-q$. Thus among p and q one number is odd, the other one is even. One may define $p'' := p' + q'$ and $q'' := p' - q'$, which are both odd and obtain,—to be more happy without the factor $f = 2$ and prove the claim*

$$a' = f(p^2 - q^2) = 2p''q'' \,,\ b' = 2fpq = (p''^2 - q''^2)\text{ and}$$
$$c' = f(p^2 + q^2) = (p''^2 + q''^2)$$
$$(b' + ia') = (p'' + iq'')^2 = [(1+i)(p' - iq')]^2$$
$$a' + ib' = i[(1-i)(p' + iq')]^2$$
$$a + ib = i[(1-i)(dp + idq)]^2$$

□

6.3 Towards Fermat's Last Theorem

Proposition 57. *There exists no solution of*

(6.3.1) $\qquad x^4 + y^4 = z^2 \,,\ \gcd(x,y,z) = 1$

with positive integers x, y, z.

Proof. If there would exist a solution of equation (6.3.1), one among them would have the minimal value for z. We arrive at a contradiction by constructing from any solution another one with smaller value of z.

Following Theorem 20 one number among x^2 and y^2 is odd and the other one is even. Supposing that x is odd, we obtain

(6.3.2) $\qquad x^2 = p^2 - q^2 \,,\ y^2 = 2pq\ \text{and}\ z = p^2 + q^2$

where $p \geq 1$ and $1 \leq q < p$ are integers such that $\gcd(p,q) = 1$ and $2 \nmid p - q$.

From $x^2 + q^2 = p^2$ we get another primitive Pythagorean triple. Indeed

$$\gcd(x, q, p) \mid \gcd(p, q) = 1$$

Applying Theorem 20 a second time, we get the existence of u and v such that

(6.3.3) $\qquad x = u^2 - v^2 \,,\ q = 2uv\ \text{and}\ p = u^2 + v^2$

Here $u \geq 1$ and $1 \leq v < u$ are integers such that $\gcd(u, v) = 1$ and $2 \nmid u - v$.

6.3. TOWARDS FERMAT'S LAST THEOREM

Going back to the second equation from (6.3.2) we get

$$y^2 = 2pq = 4puv \text{ and hence } \left(\frac{y}{2}\right)^2 = puv$$

Since p, u, v are <u>pairwise</u> relatively prime, by lemma 62 there exist positive integers a, b, c such that

$$u = a^2, \ v = b^2, \ p = c^2$$

Finally the third equation from (6.3.3) yields

$$a^4 + b^4 = c^2$$

We need still to check the inequality $c < z$. Indeed

$$c^4 = p^2 < p^2 + q^2 = z \leq z^4$$

Thus we have constructed from any solution of (6.3.1) one with smaller $z' = c$. This is impossible, and hence equation (6.3.1) has no solution. $\square$

Lemma 66. *For any solution of*

(6.3.4) $\qquad x^4 - y^4 = z^2, \ \gcd(x, y, z) = 1$

with positive integers x, y, z and y even, there exists a solution with smaller value for x.

Proof. Following Theorem 20 for $z^2 + y^4 = x^4$ one number among x^2 and y^2 is odd and the other one is even. Supposing that y is even, we get that x and z are odd.

Remark. Hence $x^2 \equiv 1 \pmod{8}$, $x^4 \equiv 1 \pmod{16}$, $z^2 = x^4 - y^4 \equiv 1 \pmod{16}$. Finally $z \equiv \pm 1 \pmod 8$. Not enough

From Theorem 20 we obtain

(6.3.5) $\qquad z = p^2 - q^2, \ y^2 = 2pq \text{ and } x^2 = p^2 + q^2$

where $p \geq 1$ and $1 \leq q < p$ are integers such that $\gcd(p, q) = 1$ and $2 \nmid p - q$.

From $p^2 + q^2 = x^2$ we get another primitive Pythagorean triple. Indeed $\gcd(p, q, x) \mid \gcd(p, q) = 1$. Applying Theorem 20 a second time, we get the existence of u and v such that

(6.3.6) $\qquad p = u^2 - v^2, \ q = 2uv \text{ and } x = u^2 + v^2$

Here $u \geq 1$ and $1 \leq v < u$ are integers such that $\gcd(u,v) = 1$ and $2 \nmid u - v$.

Going back to the second equation from (6.3.5) we get

$$y^2 = 2pq = 4puv \text{ and hence } \left(\frac{y}{2}\right)^2 = puv$$

Since p, u, v are pairwise relatively prime, by lemma 62 there exist positive integers a, b, c such that

$$u = a^2, \ v = b^2, \ p = c^2$$

Finally the first equation from (6.3.6) yields

$$p = u^2 - v^2 \text{ and hence } a^4 - b^4 = c^2$$

We need still to check the inequality $a < x$ holds. Indeed

$$a \leq a^4 = b^4 + c^2 = v^2 + p = u^2 < u^2 + v^2 = x$$

Thus we have constructed from any solution of (6.3.4) one with smaller $x' = a$. $\square$

Lemma 67. *For any solution of*

(6.3.4) $\qquad x^4 - y^4 = z^2, \ \gcd(x, y, z) = 1$

with positive integers x, y, z and y odd, there exists a solution with smaller value for x.

Proof. Following Theorem 20 we obtain from $y^4 + z^2 = x^4$

(6.3.7) $\quad y^2 = p^2 - q^2 = (p-q)(p+q), \ z = 2pq \text{ and } x^2 = p^2 + q^2$

where $p \geq 1$ and $1 \leq q < p$ are integers such that $\gcd(p,q) = 1$ and $2 \nmid p - q$. It is now astonishingly simple:

$$p^4 - q^4 = (p^2 - q^2)(p^2 + q^2) = (yx)^2$$

is a new solution of (6.3.4). Is it indeed smaller since

$$p^2 = x^2 - q^2 < x^2$$

implies $x' = p < x$. Thus we have constructed from any solution of (6.3.4) one with smaller $x' = p < x$. $\square$

6.3. TOWARDS FERMAT'S LAST THEOREM

Proposition 58. *There exists no solution of*

(6.3.4) $$x^4 - y^4 = z^2 , \; \gcd(x,y,z) = 1$$

with positive integers x, y, z.

Proof. If there would exist a solution of equation (6.3.4), one among them would have the minimal value for x. We arrive at a contradiction by constructing from any solution another one with smaller value of x. This has been done in lemma 66 in the case that y is even, and in lemma 67 in the case that y is odd. Hence equation (6.3.4) has no solution. □

Corollary 27. *In any Pythagorean triple $a^2 + b^2 = c^2$ at most one number among a, b, c can be a perfect square.*

Corollary 28 (Fermat). *There exists no solution of $x^4 + y^4 = y^4$ with nonzero integers x, y, z.*

Problem 91. *Find all primitive Pythagorean triples $a^2 + b^2 = c^2$ for which c is a perfect square.*

Answer. By assume $c = d^2$. Hence one gets from equation (6.2.1)

$$\pm a = p^2 - q^2 , \; b = 2pq \text{ and } d^2 = p^2 + q^2$$

where $p \geq 1$ and $q \geq 1$ are integers such that $\gcd(p,q) = 1$ and p is odd, q is even. Because of the third equation p, q, d is a primitive Pythagorean triple. Applying equation (6.2.1) once more we conclude

$$p = u^2 - v^2 , \; q = 2uv \text{ and } d = u^2 + v^2$$

where $u > 1$ and $1 \leq v < u$ are integers such that $\gcd(u,v) = 1$ and $2 \nmid u - v$. Altogether we have obtained

$$\pm a = (u^2 - v^2)^2 - 4u^2v^2 = (u^2 - v^2 - 2uv)(u^2 - v^2 + 2uv),$$
$$b = 4(u^2 - v^2)uv \text{ and } c = (u^2 - v^2)^2 + 4u^2v^2 = (u^2 + v^2)^2$$

where $u > 1$ and $1 \leq v < u$ are integers such that $\gcd(u,v) = 1$ and $2 \nmid u - v$. The smallest example is

$$u = 2, \; v = 1 \quad -a = 9 - 19 = -7, \; b = 24, \; c = 25$$

Chapter 7

Constructible Polygons

7.1 Fermat Primes

7.1.1 A Short Paragraph about Fermat Numbers

Definition 34 (Fermat number, Fermat prime). The numbers
$$F_n = 2^{2^n} + 1$$
for any integer $n \geq 0$ are called *Fermat numbers*. A Fermat number which is prime is called a *Fermat prime*.

Lemma 68 (Fermat). *If p is any odd prime, and $p-1$ is a power of two, then $p = F_n$ is a Fermat prime.*

Proof. Assume $p = 1 + 2^a$ and $a \geq 3$ odd. Because of the factorization $a = 2^n \cdot (2b+1)$ one can factor
$$p = 1 + 2^{2^n \cdot (2b+1)} = (1 + 2^{2^n})(1 - 2^{2^n} + 2^{2^n \cdot 2} - \cdots + 2^{2^n \cdot 2b})$$

If p is an odd prime, the only possibility is $b = 0$. Hence the exponent a cannot have any odd prime factor. Hence we are left with $p = F_n$ as the only possibility. □

The only known Fermat primes are F_n for $n = 0, 1, 2, 3, 4$. They are $3, 5, 17, 257$, and 65537, see[1].

Fermat numbers and Fermat primes were first studied by Pierre de Fermat, who conjectured 1654 in a letter to Blaise Pascal that

---
[1] sequence A019434 in OEIS

all Fermat numbers are prime, but told he had not been able to find a proof. Indeed, the first five Fermat numbers are easily shown to be prime. However, Fermat's conjecture was refuted by Leonhard Euler in 1732 when he showed, as one of his first number theoretic discoveries:

$$F_5 = 2^{2^5} + 1 = 2^{32} + 1 = 4294967297 = 641 \cdot 6700417.$$

Later, Euler proved that every prime factor p of F_n must have the form $p = i \cdot 2^{n+1} + 1$. Almost hundred years later, Lucas showed that even $p = j \cdot 2^{n+2} + 1$. These results give a vague hope that one could factor Fermat numbers and perhaps settle Fermat's conjecture to the negative.

There are no other known Fermat primes F_n with $n > 4$. However, little is known about Fermat numbers with large n. In fact, each of the following is an open problem:

1. Is F_n composite for all $n > 4$? By this *anti-Fermat hypothesis* $3, 5, 17, 257$, and 65537 would be the only Fermat primes.

2. Are there infinitely many Fermat primes? (Eisenstein 1844)

3. Are there infinitely many composite Fermat numbers?

4. Are all Fermat numbers square free?

As of 2012, the next twenty-eight Fermat numbers, F_5 through F_{32}, are known to be composite. As of February 2012, only F_0 to F_{11} have been completely factored. For complete information, see Fermat factoring status by Wilfrid Keller, on the internet at

http://www.prothsearch.net/fermat.html

Main Theorem 1 (Gauss-Wantzel Theorem). *A regular polygon with n sides is constructible if and only if*

$$n = 2^h p_1 \cdot p_2 \cdots p_s$$

where $p_1 \cdots p_s$ is a product of different *Fermat primes.*

If the anti-Fermat hypothesis *is true, there are exactly five Fermat primes, and hence exactly* 31 *regular constructible polygons with an odd number of sides.*

Problem 92. *Prove that the totient function $\phi(n)$ is a power of two if and only if*

$$n = 2^r \cdot \prod F_i$$

7.1. FERMAT PRIMES

where the F_i are different Fermat primes. Hence a regular polygon with n sides is constructible if and only if the totient function $\phi(n)$ is a power of two.

Problem 93. *Determine all solutions of $\phi(n) = 1\,024$. There are 12 solutions. Prime factor these solution. Why do we see from the Gauss-Wantzel Theorem that the corresponding regular n-gons are constructible with compass and straightedge.*

Answer. The natural numbers for which $\phi(n) = 1\,024$. are

$$(1285,\ 2048,\ 2056,\ 2176,\ 2560,\ 2570,$$
$$2720,\ 3072,\ 3084,\ 3264,\ 3840,\ 4080)$$

Here they are prime factored

$$1285 = 5 \cdot 257$$
$$2048 = 2^{11}$$
$$2056 = 2^3 \cdot 257$$
$$2176 = 2^7 \cdot 17$$
$$2560 = 2^9 \cdot 5$$
$$2570 = 2 \cdot 5 \cdot 257$$
$$2720 = 2^5 \cdot 5 \cdot 17$$
$$3072 = 2^{10} \cdot 3$$
$$3084 = 2^2 \cdot 3 \cdot 257$$
$$3264 = 2^6 \cdot 3 \cdot 17$$
$$3840 = 2^8 \cdot 3 \cdot 5$$
$$4080 = 2^4 \cdot 3 \cdot 5 \cdot 17$$

These number are all products of some power of two with different Fermat primes. By the Gauss-Wantzel Theorem, the corresponding regular n-gons are constructible with compass and straightedge. □

Corollary 29 (Gauss-Wantzel). *For all natural $n \geq 3$ are equivalent*

(a) *The regular polygon with n sides is constructible with compass and straightedge.*

(b) *The totient function $\phi(n)$ is a power of two.*

(c) $n = 2^h p_1 \cdot p_2 \cdots p_s$ where $p_2 \cdots p_s$ is a product of different Fermat primes.

(a) *implies* (b). We assume that the regular n-gon is constructible with compass and straightedge. Hence the roots of the polynomial equation $\Phi_n(z) = 1$ are in a field extension of the rationals the dimension of which is a power of two. A full proof of necessity was given by Pierre Wantzel in 1837. (More details are given in Hartshorn's book *Euclid and Beyond*.) Hence the totient function $\phi(n) = \deg \Phi_n$ is a power of two. □

(b) *implies* (c). We assume the totient function $\phi(n)$ is a power of two. Hence the prime decomposition of n is a power of two multiplied with simple odd prime factors F_k for which $\phi(F_k)$ is a power of two. Hence $F_k = 1 + 2^s$ and by lemma 68 the number F_k is a Fermat prime. □

(c) *implies* (a). We assume that n has the prime decomposition

$$n = 2^h p_1 \cdot p_2 \cdots p_s$$

where the primes p_k are Fermat primes. The construtibility of the regular n-gon follows from the constructibility of the F_k-gons for the Fermat primes, which is Gauss' main result (see corollary 30 below), and some further simple transfers and bisection of angles. These steps are possible for any given angle with compass and straightedge. □

	n	$factored$
1	3	F_0
2	5	F_1
3	15	$3 \cdot 5$
4	17	F_2
5	51	$3 \cdot 17$
6	85	$5 \cdot 17$
7	255	$3 \cdot 5 \cdot 17$
8	257	F_3
9	771	$3 \cdot 257$
10	1 285	$5 \cdot 257$
11	3 855	$3 \cdot 5 \cdot 257$
12	4 369	$17 \cdot 257$
13	13 107	$3 \cdot 17 \cdot 257$
14	21 845	$5 \cdot 17 \cdot 257$
15	65 535	$3 \cdot 5 \cdot 17 \cdot 257$
16	65 537	F_4

7.1. FERMAT PRIMES

	n	factored
17	196 611	$3 \cdot 65\,537$
18	327 685	$5 \cdot 65\,537$
19	983 055	$3 \cdot 5 \cdot 65\,537$
20	1 114 129	$17 \cdot 65\,537$
21	3 342 387	$3 \cdot 17 \cdot 65\,537$
22	5 570 645	$5 \cdot 17 \cdot 65\,537$
23	16 711 935	$3 \cdot 5 \cdot 17 \cdot 65\,537$
24	16 843 009	$257 \cdot 65\,537$
25	50 529 027	$3 \cdot 257 \cdot 65\,537$
26	84 215 045	$5 \cdot 257 \cdot 65\,537$
27	252 645 135	$3 \cdot 5 \cdot 257 \cdot 65\,537$
28	286 331 153	$17 \cdot 257 \cdot 65\,537$
29	858 993 459	$3 \cdot 17 \cdot 257 \cdot 65\,537$
30	1 431 655 765	$5 \cdot 17 \cdot 257 \cdot 65\,537$
31	4 294 967 295	$3 \cdot 5 \cdot 17 \cdot 257 \cdot 65\,537$

7.1.2 The 17-gon Construction

The regular heptadecagon is a constructible polygon (that is, one that can be constructed using a compass and unmarked straightedge), as was shown by Carl Friedrich Gauss in 1796 at the age of nineteen. This proof represented the first progress in regular polygon construction in over 2000 years. It can be found in his *Disquisitiones Arithmeticae* and is reprinted in his Werke (1870-77), vol. I. Gauss's proof of the constructibility of the relies on the fact that constructibility is equivalent to expressibility of the trigonometric functions of the angle 360°/17 in terms of arithmetic operations and square root extractions.

We use radian measure for angles in this section. According to theorems from geometry the number $2\cos(2\pi/17)$ is constructible if and only if there exists a tower of dimension extensions all of dimension two

$$\mathbf{Q} = \mathbf{F}_0 \subset \mathbf{F}_1 \subset \mathbf{F}_2 \subset \cdots \subset \mathbf{F}_n \ni 2\cos\frac{2\pi}{17}$$

Since 17 is a prime the polynomial

$$\Phi_{17}(z) = \frac{z^{17} - 1}{z - 1}$$

is a cyclotomic polynomial, and is irreducible over the integers. This fact is easy to confirm with the Eisenstein criterium. Because of the irreducibility, the extension is generated by one element. Its dimension is

$$\left[\mathbf{Q}\left(\exp\frac{2\pi i}{17}\right):\mathbf{Q}\right]=\deg\Phi_{17}=16$$

We give an explicit construction of the tower with three two dimensional extensions

(7.1.1) $$\mathbf{Q}=\mathbf{F}_0\subset\mathbf{F}_1\subset\mathbf{F}_2\subset\mathbf{Q}\left(\cos\frac{2\pi}{17}\right)$$

One further quadratic extension is needed to adjoin the imaginary parts $i\sin\frac{2\pi}{17}$, the calculation of which uses simply $\sin\alpha = \pm\sqrt{1-\cos^2\alpha}$.

Here I explain my own semi empirical approach which I call the "pairing method". In this way, one avoids the primitive roots from number theory. Define

$$c_j = 2\cos\frac{2\pi j}{17}$$

Clearly c_j as a function of the integer j is symmetric and has period 17. We get the roots in question from $j = 1\ldots 8$. Moreover the addition theorem for cosine implies

(7.1.2) $$2\cos\alpha\cos\beta = \cos(\alpha+\beta) + \cos(\alpha-\beta)$$
$$c_i \cdot c_j = c_{i+j} + c_{i-j}$$

The tower (7.1.1) in now reconstructed backwards. The third step in the tower (7.1.1) constructs the c_i from four sums of pairs of the c_i. How can one find these four pairs? To get two roots c_i and c_j by means of a quadratic equation, we need to know both the sum $c_i + c_j$ and the product $c_i c_j = c_{i+j} + c_{i-j}$. To have the product available, the four pairs have to be chosen according to the following rules:

> If c_i and c_j are paired, then c_{i+j} and c_{i-j} are paired, too.

> If c_i and c_j are paired, then c_{2i} and c_{2j} are paired, too.

Problem 94. *Convince yourself that the only way to get four pairs from the numbers* $\{1,2,3,4,5,6,7,8\}$ *is*

$$\{c_1, c_4\}\quad \{c_2, c_8\}\quad \{c_3, c_5\}\quad \{c_6, c_7\}$$

7.1. FERMAT PRIMES

Solution. We see that 1 and 2 cannot be paired. Otherwise we would get the 2, 4 which is contradictory to the rule. Can 1, 3 be a pair? No, one would get the further pairs 2, 4 and 2, 6,—again contradictory to the rule. The next case 1, 4 leads to the above success. □

Now we guess that

$$\mathbf{F}_2 = \mathbf{Q}(c_1 + c_4) = \mathbf{Q}(c_2 + c_8) = \mathbf{Q}(c_3 + c_5) = \mathbf{Q}(c_6 + c_7)$$

To get the previous extension $\mathbf{F}_2/\mathbf{F}_1$, we need to find out which two pairs have to be added.
A bid of calculation shows

$$(c_1 + c_4)(c_2 + c_8) = \sum_{i=1}^{8} c_i \quad \text{and} \quad (c_3 + c_5)(c_6 + c_7) = \sum_{i=1}^{8} c_i$$

whereas in other such products some c_i occur more than once. We guess that

$$\mathbf{F}_1 = \mathbf{Q}(c_1 + c_4 + c_2 + c_8) = \mathbf{Q}(c_3 + c_5 + c_6 + c_7)$$

For the first extension we need still the product

$$(c_1 + c_4 + c_2 + c_8)(c_3 + c_5 + c_6 + c_7) = 4 \sum_{i=1}^{8} c_i$$

and the total sum $\sum_{i=1}^{8} c_i$. Before going on, recapitulate what we have obtained so far:

Lemma 69. *Let c_j be any real valued function of the integer j with the properties:*

$$c_j = c_{-j} = c_{17-j} \quad \text{and} \quad c_i \cdot c_j = c_{i+j} + c_{i-j}$$

for all integer i and j. Then

$$c_1 c_4 = c_3 + c_5 \quad \text{and} \quad c_3 c_5 = c_2 + c_8$$
$$\text{and} \quad c_2 c_8 = c_6 + c_7 \quad \text{and} \quad c_6 c_7 = c_1 + c_4$$

$$(c_1 + c_4)(c_2 + c_8) = \sum_{i=1}^{8} c_i \quad \text{and} \quad (c_3 + c_5)(c_6 + c_7) = \sum_{i=1}^{8} c_i$$

$$(c_1 + c_4 + c_2 + c_8)(c_3 + c_5 + c_6 + c_7) = 4 \sum_{i=1}^{8} c_i$$

From a known $\sum_{i=1}^{8} c_i$, one can calculate all c_i with a tree of quadratic equations.

The sum of all eight c_i can be obtained from a complex geometric series

$$1 + \sum_{i=1}^{8} c_i = \sum_{k=0}^{16} \exp \frac{2\pi i\, k}{17} = \frac{\exp \frac{2\pi i\, 17}{17} - 1}{\exp \frac{2\pi i}{17} - 1} = 0$$

$$\text{hence} \quad \sum_{i=1}^{8} c_i = -1$$

We can now write down the quadratic equations for the extensions. We check numerically which sign of the square root in their solutions corresponds to which partial sum of c_i.

1. The polynomial with the roots

$$x_1 = c_1 + c_4 + c_2 + c_8 \quad \text{and} \quad x_2 = c_3 + c_5 + c_6 + c_7 \quad \text{is}$$
$$(x - x_1)(x - x_2) = x^2 + x - 4 \quad \text{and the roots are}$$
$$x_1 = \frac{-1 + \sqrt{17}}{2} \quad \text{and} \quad x_2 = \frac{-1 - \sqrt{17}}{2}$$

We have to check numerically that the signs of $\sqrt{17}$ occur in that arrangement.

2. The polynomial with the roots

$$y_1 = c_1 + c_4 \quad \text{and} \quad y_2 = c_2 + c_8 \quad \text{is}$$
$$(y - y_1)(y - y_2) = y^2 - x_1 y - 1 \quad \text{and the roots are}$$
$$y_{1,2} = \frac{x_1 \pm \sqrt{x_1^2 + 4}}{2} \quad \text{hence}$$
$$y_1 = \frac{-1 + \sqrt{17} + \sqrt{34 - 2\sqrt{17}}}{4} \quad \text{and}$$
$$y_2 = \frac{-1 + \sqrt{17} - \sqrt{34 - 2\sqrt{17}}}{4}$$

We have to check numerically that the signs of the longer root occur in that arrangement.

7.1. FERMAT PRIMES

3. The polynomial with the roots
$$y_3 = c_3 + c_5 \quad \text{and} \quad y_4 = c_6 + c_7 \quad \text{is}$$
$(y - y_3)(y - y_4) = y^2 - x_2 y - 1$ and the roots are
$$y_{3,4} = \frac{x_2 \pm \sqrt{x_2^2 + 4}}{2} \quad \text{hence}$$
$$y_3 = \frac{-1 - \sqrt{17} + \sqrt{34 + 2\sqrt{17}}}{4} \quad \text{and}$$
$$y_4 = \frac{-1 - \sqrt{17} - \sqrt{34 + 2\sqrt{17}}}{4}$$

We have to check numerically that the signs of the longer root occur in that arrangement.

4. The polynomial with the roots
$$z_1 = c_1 \quad \text{and} \quad z_2 = c_4 \quad \text{is}$$
$(z - z_1)(z - z_2) = z^2 - y_1 z + y_3$ and the roots are
$$z_{1,2} = \frac{y_1 \pm \sqrt{y_1^2 - 4 y_3}}{2}$$

We have to check numerically that the signs occur in that arrangement.

5. The polynomial with the roots
$$z_3 = c_2 \quad \text{and} \quad z_4 = c_8 \quad \text{is}$$
$(z - z_3)(z - z_4) = z^2 - y_2 z + y_4$ and the roots are
$$z_{3,4} = \frac{y_2 \pm \sqrt{y_2^2 - 4 y_4}}{2}$$

We have to check numerically that the signs occur in that arrangement.

6. The polynomial with the roots
$$z_5 = c_3 \quad \text{and} \quad z_6 = c_5 \quad \text{is}$$
$(z - z_5)(z - z_6) = z^2 - y_3 z + y_2$ and the roots are
$$z_{1,2} = \frac{y_3 \pm \sqrt{y_3^2 - 4 y_2}}{2}$$

We have to check numerically that the signs occur in that arrangement.

7. The polynomial with the roots

$$z_7 = c_6 \quad \text{and} \quad z_8 = c_7 \quad \text{is}$$
$$(z - z_7)(z - z_8) = z^2 - y_4 z + y_1 \quad \text{and the roots are}$$
$$z_{3,4} = \frac{y_4 \pm \sqrt{y_4^2 - 4y_1}}{2}$$

We have to check numerically that the signs occur in that arrangement.

For the root z_1, I have finished the calculation and have obtained

$$16 \cos \frac{2\pi}{17} = -1 + \sqrt{17} + \sqrt{34 - 2\sqrt{17}} + \sqrt{w} \quad \text{with } w =$$
$$\left(-1 + \sqrt{17} + \sqrt{34 - 2\sqrt{17}}\right)^2 + 16 + 16\sqrt{17} - 16\sqrt{34 + 2\sqrt{17}}$$

Additionally to the above obtained result, computations with mathematica confirm the following result

Proposition 59 (The 17-gon with mathematica). *With $\pm l = 1 \ldots 8$ and $s_1, s_2, s_3 = \pm 1$, we obtain all eight values*

$$16 \cos \frac{2\pi l}{17} = -1 + s_1 \sqrt{17} + s_2 \sqrt{34 - 2s_1 \sqrt{17}}$$
$$+ s_2 s_3 \sqrt{68 + 12 s_1 \sqrt{17} - 2 s_2 \sqrt{34 - 2s_1 \sqrt{17}} + A + B}$$
$$\text{with } A := +2 s_1 s_2 \sqrt{17(34 - 2s_1 \sqrt{17})} \text{ and}$$
$$B := -16 s_1 s_2 \sqrt{34 + 2 s_1 \sqrt{17}}$$

in the following order

l	s_1	s_2	s_3	$(kji)_2$	$3^{4k+2j+i} \pmod{17}$
1	+1	+1	+1	0	1
-15	+1	-1	-1	$110_2 = 6$	15
3	-1	+1	+1	1	3
-13	+1	+1	-1	$100_2 = 4$	13
5	-1	+1	-1	$101_2 = 5$	5
-11	-1	-1	-1	$111_2 = 7$	11
-10	-1	-1	+1	$10_2 = 3$	10
-9	+1	-1	+1	$10_2 = 2$	9

7.1. FERMAT PRIMES

This order follows the rule
(7.1.3)
$$|l| = 3^{4k+2j+i} \pmod{17} \quad \text{with } k = \frac{1-s_3}{2}, j = \frac{1-s_2}{2}, i = \frac{1-s_1}{2}$$

Again with l from the rule (7.1.3), there hold the following formulas given by Gauss' book Disquisitiones Arithmeticae—*in modern notation:*

(7.1.4) $\quad 16\cos\dfrac{2\pi l}{17} = -1 + s_1\sqrt{17} + s_2\sqrt{34 - 2s_1\sqrt{17}} + 2s_2s_3\sqrt{w}$

$w = 17 + 3s_1\sqrt{17} - s_2\sqrt{34 - 2s_1\sqrt{17}} - 2s_1s_2\sqrt{34 + 2s_1\sqrt{17}}$

as well as the following additional less known formula

(7.1.5) $\quad 16\cos\dfrac{2\pi l}{17} = -1 + s_1\sqrt{17} + s_2\sqrt{34 - 2s_1\sqrt{17}}$

$\qquad + 2s_2s_3\sqrt{17 + 3s_1\sqrt{17} - s_1s_2\sqrt{2(85 + 19s_1\sqrt{17})}}$

Checks for proposition 59. This is a numerical check with mathematica, as well as a way to obtain formulas (7.1.4) and (7.1.5).

```
In[1]:= n = 2; Fermat = Function[n, 2^(2^n) + 1];
   Fermat[n]
Out[1]= 17

In[2]:= Table[4 k + 2 j + i, {k, 0, 1},
          {j, 0, 1}, {i, 0, 1}]
Out[2]= {{{0, 1}, {2, 3}}, {{4, 5}, {6, 7}}}

In[3]:= cos16 = -1 + Sqrt[17] s1 +
   Sqrt[34 - 2 Sqrt[17] s1] s2 +
   s2 *s3 \[Sqrt](68 + 12 Sqrt[17] s1
      - 2 Sqrt[34 - 2 Sqrt[17] s1] s2 +
      2 Sqrt[17] s1 Sqrt[34 - 2 Sqrt[17] s1] s2 -
      16 s1 Sqrt[34 + 2 Sqrt[17] s1] s2) ;

In[4]:= Table[mysigns
   = {s1 -> 1 - 2 i, s2 -> 1 - 2 j, s3 -> 1 - 2 k};
   {s1, s2, s3,
   N[(cos16 - 16
      Cos[2 Pi*Mod[3^(4 k + 2 j + i), 17]/17])]}
   /. mysigns,  {k, 0, 1}, {j, 0, 1}, {i, 0, 1}]
```

```
Out[4]= {{{{1, 1, 1, -3.55271*10^-15}, {-1, 1, 1, 0.}},
   {{1, -1, 1, 0.}, {-1, -1, 1, -1.77636*10^-15}}},
          {{{1, 1, -1,   8.88178*10^-16},
            {-1, 1, -1, -8.88178*10^-16}},
           {{1, -1, -1, 3.55271*10^-15},
            {-1, -1, -1, 8.88178*10^-16}}}}

In[5]:= cos16g = -1 + s1 Sqrt[17] +
   s2 Sqrt[34 - 2 s1 Sqrt[17]] +
   2 *s2*s3 \[Sqrt](17 + 3 s1 Sqrt[17] -
      s2 Sqrt[34 - 2 s1 Sqrt[17]]
      - 2 s1*s2 Sqrt[34 + 2 s1 Sqrt[17]]);

In[6]:= Table[mysigns = {s1 -> 1 - 2 i,
   s2 -> 1 - 2 j, s3 -> 1 - 2 k}; {s1, s2, s3,
   N[(cos16g -
   16 Cos[2 Pi*Mod[3^(4 k + 2 j + i), 17]/17])]}
   /. mysigns, {k, 0, 1}, {j, 0, 1}, {i, 0, 1}]

Out[6]= {{{{1, 1, 1, -1.77636*10^-15},
   {-1, 1, 1, 0.}},{{1, -1, 1, 0.},
   {-1, -1, 1, -1.77636*10^-15}}},
   {{{1, 1, -1, 0.}, {-1, 1, -1, -8.88178*10^-16}}
   {{1, -1, -1, 0.}, {-1, -1, -1,  8.88178*10^-16}}}}

In[7]:= cos16new = -1 + s1 Sqrt[17] +
  s2 Sqrt[34- 2 s1 Sqrt[17]] +
  2 s2*s3 \[Sqrt](17 + 3 s1 Sqrt[17]
  - s2*s1 Sqrt[2 (85 + 19 s1 Sqrt[17])]);

In[8]:= Table[mysigns = {s1 -> 1 - 2 i,
   s2 -> 1 - 2 j, s3 -> 1 - 2 k}; {s1, s2, s3,
   N[(cos16new -
   16 Cos[2 Pi*Mod[3^(4 k + 2 j + i), 17]/17])]}
   /. mysigns, {k, 0, 1}, {j, 0, 1}, {i, 0, 1}]

Out[8]= {{{{1, 1, 1, -1.77636*10^-15},
   {-1, 1, 1, 0.}},{{1, -1, 1, 0.},
   {-1, -1, 1, 0.}}}, {{{1, 1, -1, 0.},
      {-1, 1, -1, -1.77636*10^-15}}, {{1, -1, -1, 0.},
      {-1, -1, -1, -8.88178*10^-16}}}}
```

7.1. FERMAT PRIMES 181

```
In[9]:= Solve[Sqrt[17] s1 Sqrt[34 - 2 Sqrt[17] s1] -
   4 s1 Sqrt[34 + 2 Sqrt[17] s1]
        == -Sqrt[34 - 2 Sqrt[17] s1], {s1}]
Out[9]= {{s1 -> -1}, {s1 -> 1}}

In[10]:= RootReduce[(34 + 6 Sqrt[17] -
  Sqrt[34 - 2 Sqrt[17]] +
   Sqrt[17 (34 - 2 Sqrt[17])] -
       8 Sqrt[34 + 2 Sqrt[17]])/8]

Out[10]=
 Root[17 - 85 #1 + 68 #1^2 - 17 #1^3 + #1^4 &, 3]

In[11]:= MinimalPolynomial[
  Root[17 - 85 #1 + 68 #1^2 -
       17 #1^3 + #1^4 &, 3], x]
Out[11]= 17 - 85 x + 68 x^2 - 17 x^3 + x^4

In[12]:= Solve[17 - 85 x + 68 x^2 -
            17 x^3 + x^4 == 0, x]
Out[12]= {{x -> 1/4 (17 - 3 Sqrt[17]
                    - Sqrt[2 (85 - 19 Sqrt[17])])},
 {x -> 1/4 (17 - 3 Sqrt[17]
     + Sqrt[2 (85 - 19 Sqrt[17])])},
 {x -> 1/4 (17 + 3 Sqrt[17] - 2 Sqrt[85/2
     + (19 Sqrt[17])/2])},
 {x -> 1/4 (17 + 3 Sqrt[17] + 2 Sqrt[85/2
     + (19 Sqrt[17])/2])}}

In[14]:= Solve[
 Sqrt[2] s1 Sqrt[85 + 19 Sqrt[17] s1] ==
  Sqrt[34 - 2 Sqrt[17] s1]
        + 2 s1 Sqrt[34 + 2 Sqrt[17] s1], {s1}]
Out[14]= {{s1 -> -1}, {s1 -> 1}}
```

□

Remark. The primitive root 3 has not been used,— and even is not need to be known—. I have no good explanation for formula (7.1.3). It seems to be an accidental coincidence that the above procedure agrees with Gauss' procedure using 3 as a primitive root.

Problem 95. *Check that the rule (7.1.3) holds similarly for all primitive roots of 17, but of course only after modification of the meaning of s_i.*

Problem 96. *Show by symbolic manipulation that indeed*

$$s_1\sqrt{17}\sqrt{34 - 2s_1\sqrt{17}} - 4s_1\sqrt{34 + 2s_1\sqrt{17}} = -\sqrt{34 - 2s_1\sqrt{17}}$$

holds for $s_1 = \pm 1$.

Problem 97. *Show by symbolic manipulation that indeed*

$$s_1\sqrt{2(85 + 19s_1\sqrt{17})} = \sqrt{34 - 2s_1\sqrt{17}} + 2s_1\sqrt{34 + 2s_1\sqrt{17}}$$

holds for $s_1 = \pm 1$.

Of course, all these empirical checks of signs are still unsatisfactory. More important, one wants to set up a more systematic procedure at the very beginning. It is the merit of Gauss to have found satisfactory answers to the latter problem. Thus he opens up the possibility to construct the polygons for other Fermat primes.

The signs $s_1, s_2, s_3 = \pm$ correspond to the digits $i, j, k = 0$ or 1. We refer to the binary tree from the proof of Gauss' theorem in the next section. While building the tree from its root, the signs s_1, s_2, s_3 determine the branches and the signs $s_.$ are combined to the binary number $(k\,j\,i)_2 = 4k + 2j + i \in \mathcal{B}$.

The list of the seventeen remainders of 3^k modulo 17 for $k = 0, 1, \ldots, 15$ is

$$[\,1, 3, -8, -7\,,\ -4, 5, -2, -6\,,\ -1, -3, 8, 7\,,\ 4, -5, 2, 6\,]$$

We see they are all different. Hence one calls 3 a *primitive root* modulo 17. Consequently the sixteen numbers

$$d_r = \exp\left(\frac{2\pi i}{17} 3^r\right)$$

are a permutation of the roots of $\frac{z^{17}-1}{z-1} = 0$. After pairing conjugate complex roots d_r and d_{r+8}, we get the ordered list

$$[\,c_1, c_3, c_8, c_7\,,\ c_4, c_5, c_2, c_6\,] = d_r + d_{r+8} \quad \text{for } r = 0, 1, \ldots, 7$$

7.1. FERMAT PRIMES

The pairs formed at the beginning of the empirical procedure, and sums of four in the next step are

$$c_1 + c_4 = d_0 + d_4 + d_8 + d_{12}$$
$$c_2 + c_8 = d_2 + d_6 + d_{10} + d_{14}$$
$$c_3 + c_5 = d_1 + d_5 + d_9 + d_{13}$$
$$c_6 + c_7 = d_3 + d_7 + d_{11} + d_{15}$$
$$c_1 + c_4 + c_2 + c_8 = d_0 + d_2 + d_4 + d_6 + d_8 + d_{10} + d_{12} + d_{14}$$
$$c_3 + c_5 + c_6 + c_7 = d_1 + d_3 + d_5 + d_7 + d_9 + d_{11} + d_{13} + d_{15}$$

One can see from sums of d_r much more easily how to set up the tree of quadratic equations. Indeed in Proposition 59

$$l \equiv (3^r \bmod 17) \quad \text{and} \quad r = \frac{1}{2}(1 - s_1) + (1 - s_2) + 2(1 - s_3)$$

These hints must be enough towards a guess how to generalize the procedure for other Fermat primes. Results which I have obtained by putting together Gauss' ideas with numerical computation are given in the next two sections.

7.1.3 Gauss' Polygons

Let $F_n \geq 5$ be a Fermat prime. The known cases are only $5, 17, 257$ and $65\,537$. The construction of the regular F_n-gon by ruler and compass involves the solution of the equation

(7.1.6) $$\Phi_{F_n}(z) = 0$$

by means of a binary tree of quadratic equations. Of course, Euler's formula
$e^{it} = \cos t + i \sin t$ gives the analytic solutions

(7.1.7) $$z_k = \exp \frac{2k\pi i}{F_n} \quad \text{with } k = 1, 2, \ldots, F_n - 1.$$

Gauss' method proceeds by finding at first suitable sums of z_k. It turns out that these can be obtained from <u>quadratic equations</u> only. At first we remark that

$$s := \sum_{1 \leq k < F_n} z_k = -1$$

since the z_k and 1 are the roots of $z^{F_n} - 1 = 0$, and by Viëta's formula these roots have the sum zero. By corollary 34, we get the primitive root $a = 3$ for any Fermat prime $F_n \geq 5$. Hence all $k \in \mathcal{U}(F_n)$ are the powers

$$k \equiv 3^r \mod F_n \quad \text{with } 0 \leq r < F_n - 1.$$

Since F_n is a prime holds $\phi(F_n) = F_n - 1 = 2^{2^n}$. Indeed we have obtained a bijection $r \in \mathbf{Z}(\phi(F_n)) \mapsto k \in \mathcal{U}(F_n)$. To determine k mod F_n, it is enough to know r mod $\phi(F_n)$. One puts this formula into equation (7.1.7) to obtain

(7.1.8) $$d_r := \exp \frac{2\pi i \cdot 3^r}{F_n} \quad \text{with } 0 \leq r < F_n - 1.$$

Hence it is convenient to put the exponent r into binary representation

(7.1.9) $$r = \sum_{0 \leq q < 2^n} b_q 2^q \quad \text{with the binary digits } b_q \in \{0, 1\}$$

Only 2^n binary digits are needed to determine $3^r \mod F_n$. We see that

$$\mathcal{B} = \{(b_{2^n-1} \ldots b_0) : b_q \in \{0, 1\} \text{ for } 0 \leq q < 2^n \}$$

is the entire set of binary digit-tuples involved. The corresponding binary integers are the set

(7.1.10) $$\mathcal{R}(n, 0, 0) = \{ \sum_{0 \leq q < 2^n} b_q 2^q : b_q \in \{0, 1\} \text{ for } 0 \leq q < 2^n \}$$

Formula (7.1.8) is now used to determine the 2^{2^n} vertices of the regular F_n-gon different from the exceptional vertex 1. Let $1 \leq q \leq 2^n$. For any fixed lowest digits $l_{q-1} \ldots l_0$ in formula (7.1.9) the low digits yield

(7.1.11) $$l = \sum_{0 \leq i < q} l_i 2^i \quad \text{with the binary digits } l_i \in \{0, 1\}$$

which is a number $0 \leq l < 2^q$. We define the subsets $\mathcal{S}(n, q, l) \subseteq \mathcal{U}(F_n)$ as well as sums of corresponding polygon vertices $S(n, q, l) \in \mathbb{C}$ by setting

$$\mathcal{S}(n, q, l) = \{3^r \in \mathcal{U}(F_n) : r = \sum_{0 \leq i < q} l_i 2^i + \sum_{q \leq j < 2^n} h_j 2^j$$
$$\text{and } h_j \in \{0, 1\} \}$$

7.1. FERMAT PRIMES

(7.1.12) $\quad \mathcal{R}(n,q,l) = \{r : r = l + x2^q \text{ with } 0 \leq x < 2^{2^n - q}\}$

(7.1.13) $\quad \mathcal{S}(n,q,l) = \{3^r \in \mathcal{U}(F_n) : r \in \mathcal{R}(n,q,l)\}$

(7.1.14) $\quad S(n,q,l) = \sum \{\exp \dfrac{2\pi i \cdot k}{F_n} : k \in \mathcal{S}(n,q,l)\}$

At each vertex of the rooted binary tree with 2^n levels of descendents is put a suitable sum $S(n,q,l)$. Here $1 \leq q \leq 2^n$ is the level in the tree and $0 \leq l < 2^q$ enumerates the vertices in a given level, say from left to right. At each tree vertex, the low digits (7.1.11) are fixed. The root of the binary tree has level 0. At the root is put the sum $s = -1$, which is the sum of all d_r.

At the vertices of the q-th level of the tree are put those sums for which the q lowest digits $l_{q-1} \ldots l_0$ are fixed, and the sum is taken over all choices for the high digits $h_{2^n - 1} \ldots h_q$. The node $[l_{q-1} \ldots l_0]$ gets the two children $[1, l_{q-1} \ldots l_0]$ and $[0, l_{q-1} \ldots l_0]$. Increasing the level by one, one more low digit becomes fixed, whereas the sums over the high digits get shorter by one digit. The numbers (7.1.8) are obtained at the 2^{2^n} leaves of the tree.

The sums $S(n,q,l)$ are needed during this process. They may be computed by means of quadratic equations. How does one get the sums on the $q+1$-st level from those on the q-th level? To get the two children' sums from its (one!) [2] parent by means of a quadratic equation, one needs to know their sum and product. The sum is easy:

$$P := S(n, q+1, l) + S(n, q+1, l + 2^q) = S(n, q, l)$$

To determine the product

$$Q := S(n, q+1, l) \cdot S(n, q+1, l + 2^q)$$

I need god, Gauss and good guesses. Clearly, the multiplication leads to addition of exponents.

$$S(n, q+1, l) \cdot S(n, q+1, l+2^q)$$
$$= \sum_{k \in \mathcal{S}(n,q+1,l)} \exp \dfrac{2\pi i \cdot k}{F_n} \cdot \sum_{k' \in \mathcal{S}(n,q+1,l+2^q)} \exp \dfrac{2\pi i \cdot k'}{F_n}$$

(7.1.15) $\quad = \sum \{\exp \dfrac{2\pi i \cdot (k+k')}{F_n} :$
$\qquad (k,k') \in \mathcal{S}(n, q+1, l) \times \mathcal{S}(n, q+1, l+2^q)\}$

---

[2] In any rooted tree a node has always only one parent. But a parent node of a binary tree can have up to two children.

The sums $k + k'$ occurs over all pairs (k, k') from the Cartesian product set
$\mathcal{S}(n, q+1, l) \times \mathcal{S}(n, q+1, l+2^q)$. Not too convenient!

Finally the sums $S(n, q+1, l)$ and $S(n, q+1, l+2^q)$ can in the the next step be determined from the quadratic equation $x^2 - Px + Q = 0$ with the coefficients

$$P = S(n, q+1, l) + S(n, q+1, l+2^q)$$
$$Q = S(n, q+1, l) \cdot S(n, q+1, l+2^q)$$

Lemma 70. *From the root to the $2^n - 1$-st level all sums $S(n, q, l)$ turn out to be real.*

Reason. Let

(7.1.16) $$h = \frac{F_n - 1}{2} = 2^{2^n - 1}$$

Since $3^h \equiv -1 \mod F_n$, we get from equation (7.1.8) conjugate complex pairs $d_{r+h} = \overline{d_r}$ and

(7.1.17) $$d_r + d_{r+h} = 2\cos\frac{2\pi \cdot 3^r}{F_n}$$

is real for all r. Indeed, for $q < 2^n$ each set $\mathcal{S}(n, q, l)$ contains with any member 3^r a second member $3^{r+h} \equiv -3^r \mod F_n$. Since $d_{r+h} = \overline{d_r}$ are conjugate complex to each other, the sums $S(n, q, l)$ are sums of cosinus terms (7.1.17), and hence real. □

Lemma 71. $1 \leq q \leq 2^n - 1$. *To produce the $q+1$-st level sums from the q-th level sums one needs the product from equation (7.1.15). The special case $k + k' \equiv 0 \mod F_n$ occurs only for the maximal last level $q = 2^n - 1$. In this special case always hold $k + k' \equiv 0 \mod F_n$ and $S(n, q+1, l) \cdot S(n, q+1, l+2^q) = 1$.*

Reason. Let $k \in \mathcal{S}(n, q+1, l)$ and $k' \in \mathcal{S}(n, q+1, l+2^q)$. Let $k = 3^r \mod F_n$ and assume $k + k' \equiv 0 \mod F_n$. Hence $k' = 3^{r+h}$ with $h = 2^{2^n-1}$. By definition (7.1.12) $r = l + x2^{q+1}$ and $0 \leq x < 2^{2^n - q - 1}$. Similarly $r + h = l + 2^q + y2^{q+1}$ and $0 \leq y < 2^{2^n - q - 1}$.

Calculation in the double exponent are done modulo $F_n - 1 = 2^{2^n}$, in other words with n binary digits. One gets

$$2^{2^n - 1} = h \equiv 2^q + (y - x)2^{q+1} \pmod{2^{2^n}} \text{ and } q \leq 2^n - 1$$

For $q < 2^n - 1$ we get a contradiction. Hence $k + k' \equiv 0$ is impossible. For the second last level $q = 2^n - 1$, because of equation (7.1.16) the congruence holds indeed. The node sets are conjugate complex pairs. Their sum gives the double cosins (7.1.17). □

7.1. FERMAT PRIMES

I use the notation

$$\oplus A \times B := \text{Multiset}\{a + b : a \in A \text{ and } b \in B\}$$

mainly for the multiset

$$\oplus S(n, q+1, l) \times S(n, q+1, l+2^q) = \text{Multiset } \{3^r + 3^{r'} : \\ r \in \mathcal{R}(n, q+1, l) \text{ and } r' \in \mathcal{R}(n, q+1, l+2^q)\}$$

Lemma 72. *We now explain how to go to the last level. With $q = 2^n - 1$ and $h = 2^q$ we have to solve the quadratic equations $x^2 - Px + Q = 0$ with the coefficients*

$$P = S(n, 2^n - 1, l) = S(n, 2^n, l) + S(n, 2^n, l + h) = 2\cos\frac{2\pi \cdot 3^l}{F_n}$$

$$Q = S(n, 2^n, l) \cdot S(n, 2^n, l + 2^{2^n - 1}) = 1$$

As expected, one gets the solutions

$$S(n, 2^n, l), \ S(n, 2^n, l + h) = \exp\pm\frac{2\pi i \cdot 3^l}{F_n}$$

Reason. With $q = 2^n - 1$ and $2^q = h$ holds

(7.1.18)
$$S(n, 2^n - 1, l)$$
$$= \{3^r \mod F_n : r = l + xh \text{ and } 0 \le x < 2\} = \{\pm 3^l\}$$
$$S(n, 2^n - 1, l) = \sum\{\exp\frac{2\pi i \cdot k}{F_n} : k \in \{\pm 3^l\}\} = 2\cos\frac{2\pi \cdot 3^l}{F_n}$$

$$\oplus S(n, 2^n, l) \times S(n, 2^n, l + h)$$
$$= \text{Multiset } \{3^r + 3^{r'} : 3^r = 3^l \text{ and } 3^{r'} = -3^l\} = \{0\}$$
$$S(n, 2^n, l) \cdot S(n, 2^n, l + h) = 1$$

$\square$

Lemma 73. *To go to the first level one gets the equation $x^2 + x - 2^{n-2} = 0$ hence the solutions*

$$S(n, 1, 0) = \frac{-1 + \sqrt{F_n}}{2} \quad \text{and} \quad S(n, 1, 1) = \frac{-1 - \sqrt{F_n}}{2}$$

or exchanged. The signs of the roots are harder to know.

Remark. Formula (7.1.36) for $q = 0$ is

$$-\frac{F_n - 1}{4} = S(n,1,0) \cdot S(n,1,1) = e[n,0,0]S(n,0,0) = -e[n,0,0]$$

$$e[n,0,0] = 2^{2^n - 2}$$

The same result follows from lemma 80 below.

Reason. The set $\mathcal{S}(n,1,0)$ consists of the $h = 2^{2^n-1}$ quadratic residues, and the set $\mathcal{S}(n,1,1)$ of the 2^{2^n-1} quadratic non-residues. Assume that

$$a = 3^{2s} + 3^{2s'+1} \in \oplus \mathcal{S}(n,1,0) \times \mathcal{S}(n,1,1)$$

is a quadratic residue,— as for example $3^0 + 3^1 = 4$. Then $3^{2t}a = 3^{2(s+t)} + 3^{2(s'+t)+1}$ are quadratic residues for all t. By mapping $(s,s') \mapsto (s+t, s'+t)$ for $0 \le t < h$ one produces in this way h different quadratic residues. But $3^{2t-1}a = 3^{2(s'+t)} + 3^{2(s+t-1)+1}$ are quadratic non-residues for all t. By $(s,s') \mapsto (s'+t, s+t-1)$ for $0 \le t < h$ one obtains in this way h different quadratic non-residues. The entire set $\mathcal{S}(n,0,0) = \mathcal{U}(F_n)$ is covered, indeed $\frac{h}{2}$ times as one sees choosing $s' - s = 1 \ldots \frac{h}{2}$.

Remark. We claim that $3^{2s} + 3^{2s'+1}$ is a quadratic residue if $s+s'$ is even, as for example $3^0 + 3^2$. Moreover $3^{2s} + 3^{2s'+1}$ is a quadratic non-residue if $s + s'$ is odd, as for example $3^0 + 3^1$.

In other word, the product multiset $\oplus \mathcal{S}(n,1,0) \times \mathcal{S}(n,1,1)$ contains $\frac{h^2}{2}$ quadratic residues and $\frac{h^2}{2}$ quadratic non-residues and

$$\oplus \mathcal{S}(n,1,0) \times \mathcal{S}(n,1,1) = \frac{h}{2}\mathcal{S}(n,0,0) = 2^{2^n-2}\mathcal{U}(F_n)$$

The quadratic equation $x^2 - Px + Q = 1$ with the coefficients

$$P = S(n,1,0) + S(n,1,1) = S(n,0,0) = -1$$
$$Q = S(n,1,0) \cdot S(n,1,1) = -2^{2^n-2}$$

has the solutions $S(n,1,0)$, $S(n,1,1)$ given above. □

The hard part is to set up the quadratic equations for the levels $0 < q < 2^n - 1$. We do not know much more than

$$\mathcal{S}(n,0,0) = \mathcal{U}(F_n)$$

$$\#\mathcal{S}(n,q,l) = 2^{2^n-q}$$

7.1. FERMAT PRIMES

The sum $k+k'$ in formula (7.1.15) occurs over all pairs (k,k') from the Cartesian product set $\mathcal{S}(n,q+1,l) \times \mathcal{S}(n,q+1,l+2^q)$! This Cartesian product has $2^{2^n-q-1} \cdot 2^{2^n-q-1} = 2^{2(2^n-q-1)}$ members, many of which result in equal sums $k+k'$ since there are only 2^{2^n} choices.

Conjecture 4. *The first guess is that always 2^{2^n-2q-2} terms are equal and all 2^{2^n} possible values for $k+k'$ occur. But we see that is only possible for $q \leq 2^{n-1} - 1$. One could get*

$$\oplus \mathcal{S}(n,q+1,l) \times \mathcal{S}(n,q+1,l+2^q) = 2^{2^n-2q-2}\mathcal{U}(F_n)$$
$$\text{for } q \leq 2^{n-1} - 1.$$

The second guess is

$$\oplus \mathcal{S}(n,q+1,l) \times \mathcal{S}(n,q+1,l+2^q) = 2^{2^n-q-2}\mathcal{S}(n,q,l^*)$$
$$\text{for } 2^{n-1} \leq q < 2^n - 1.$$

and $l^* = (l+1) \mod 2^q$. The latter multiset has 2^{2^n-q} members, and each one occurs 2^{2^n-q-2} times.

Conjecture 5. *The third guess is that for $2^{n-1} \leq q < 2^n - 1$, always $2^{2^n-2q-2+q'}$ terms are equal and all $2^{2^n-q'}$ terms occur*

$$\oplus \mathcal{S}(n,q+1,l) \times \mathcal{S}(n,q+1,l+2^q) = 2^{2^n-2q-2+q'}S(n,q',l^*)$$
$$\text{for } 2^{n-1} \leq q < 2^n - 1.$$

and $l^* = (l+1) \mod 2^{q'}$. The latter multiset has $2^{2^n-q'}$ members, and each one occurs $2^{2^n-q-2+q'} = 1$ times.

$$q' = 2q + 2 - 2^n, \ 2 \leq q' < 2^n$$

Proposition 60 (The 17-gon is solved). *The checked results for the 17-gon from the previous section are:*

(7.1.19) $S(2,1,0) \cdot S(2,1,1) = -4$

(7.1.20) $S(2,2,0) \cdot S(2,2,2) = -1$

(7.1.21) $S(2,2,1) \cdot S(2,2,3) = -1$

(7.1.22) $S(2,3,0) \cdot S(2,3,4) = S(2,2,1)$

(7.1.23) $S(2,3,1) \cdot S(2,3,5) = S(2,2,2)$

(7.1.24) $S(2,3,2) \cdot S(2,3,6) = S(2,2,3)$

(7.1.25) $S(2,3,3) \cdot S(2,3,7) = S(2,2,0)$

This confirms both conjectures 4 and 5 to hold for $n = 2$ and $F_2 = 17$.

Proof. The results from the last section are in the present notation

$$(c_1 + c_4 + c_2 + c_8)(c_3 + c_5 + c_6 + c_7) = -4$$
$$S(2,1,0) \cdot S(2,1,1)$$
$$= (d_0 + d_2 + d_4 + d_6 + d_8 + d_{10} + d_{12} + d_{14})$$
$$\cdot (d_1 + d_3 + d_5 + d_7 + d_9 + d_{11} + d_{13} + d_{15})$$
$$= -4$$

$$(c_1 + c_4)(c_2 + c_8) = -1$$
$$S(2,2,0) \cdot S(2,2,2)$$
$$= (d_0 + d_4 + d_8 + d_{12})(d_2 + d_6 + d_{10} + d_{14}) = -1$$
$$(c_3 + c_5)(c_6 + c_7) = -1$$
$$S(2,2,1) \cdot S(2,2,3)$$

$$= (d_1 + d_5 + d_9 + d_{13})(d_3 + d_7 + d_{11} + d_{15}) = -1$$
$$c_i \cdot c_j = c_{i+j} + c_{i-j}$$
$$c_1 \cdot c_4 = c_3 + c_5$$
$$c_2 \cdot c_8 = c_6 + c_7$$
$$c_3 \cdot c_5 = c_2 + c_8$$
$$c_6 \cdot c_7 = c_1 + c_4$$

$$S(2,3,0) \cdot S(2,3,4) = (d_0 + d_8)(d_4 + d_{12}) = c_1 \cdot c_4 = c_3 + c_5$$
$$= (d_1 + d_5 + d_9 + d_{13}) = S(2,2,1)$$
$$S(2,3,1) \cdot S(2,3,5) = (d_1 + d_9)(d_5 + d_{13}) = c_3 \cdot c_5 = c_2 + c_8$$
$$= (d_2 + d_6 + d_{10} + d_{14}) = S(2,2,2)$$
$$S(2,3,2) \cdot S(2,3,6) = (d_2 + d_{10})(d_6 + d_{14}) = c_2 \cdot c_8 = c_6 + c_7$$
$$= (d_3 + d_7 + d_{11} + d_{15}) = S(2,2,3)$$
$$S(2,3,3) \cdot S(2,3,7) = (d_3 + d_{11})(d_7 + d_{15}) = c_7 \cdot c_6 = c_1 + c_4$$
$$= (d_0 + d_4 + d_8 + d_{12}) = S(2,2,0)$$

In short, we have checked

$$S(n, q+1, l) \cdot S(n, q+1, l+2^q) = -2^{2^n - 2q - 2} \quad \text{for } q \leq 2^{n-1} - 1.$$

hence $q = 0$ and $q = 1$. This turns out to be (7.1.19) (7.1.20) and (7.1.21).

7.1. FERMAT PRIMES

The second guess is that for $2^{n-1} \le q < 2^n$ hence $q = 2$,

$$S(2,3,l) \cdot S(2,3,l+4) = S(2,2,l^*) \quad \text{for } q = 2.$$

This holds with $l^* = (l+1) \mod 4$ and turns out to be (7.1.22) (7.1.23) (7.1.24) and (7.1.25). □

The above conjectures turn out to work for the 17-gon, but are not correct for the 257-gon. To save the day and prove at least some theoretical principles, I now borrow an idea from Fourier analysis.

Lemma 74. *Let $0 \le q < 2^n - 1$. Let $\mathcal{E}(n,q)$ be any finite multiset built from multiple copies of elements from $\mathcal{U}(F_n)$ such that elements 3^r and 3^s occur with equal multiplicities if $2^q \mid r - s$. Any such multiset is a linear combination*

$$\mathcal{E}(n,q) = \bigcup_{0 \le k < 2^q} f_{[q,k]} \mathcal{S}(n,q,k)$$

with non-negative coefficients $c_{[q,k]}$. Moreover these coefficients are obtained with a Fourier-typ analysis. One gets

$$2^{2^n - q} \cdot f_{[q,k]} = \#\mathcal{E}(n,q) \cap \mathcal{S}(n,q,k)$$

for all $0 \le k < 2^q$.

Proof. The linear combinations (74) exhaust the above specified multisets $\mathcal{E}(n,q)$. The orthogonality relation

$$\#\mathcal{S}(n,q,l) \cap \mathcal{S}(n,q,k) = 2^{2^n - q} \delta_{lk}$$

for $0 \le l < 2^q$ and $0 \le k < 2^q$ makes the Fourier analysis feasible. □

Lemma 75. *Let $0 \le q \le 2^n - 1$. we may extend the definition*

(7.1.12) $\quad \mathcal{R}(n,q,l) = \{r : r = l + x2^q \text{ with } 0 \le x < 2^{2^n - q}\}$

(7.1.13) $\quad \mathcal{S}(n,q,l) = \{3^r \in \mathcal{U}(F_n) : r \in \mathcal{R}(n,q,l)\}$

(7.1.14) $\quad S(n,q,l) = \sum \{\exp \dfrac{2\pi i \cdot k}{F_n} : k \in \mathcal{S}(n,q,l)\}$

to all integer l. In that case holds

$$\mathcal{R}(n,q,l+2^q) = \mathcal{R}(n,q,l) \pmod{F_n - 1}$$
$$\mathcal{S}(n,q,l+2^q) = \mathcal{S}(n,q,l)$$

Proof. Since $3^{F_n-1} \equiv 1 \mod F_n$ one may simplify

$$\begin{aligned}
\mathcal{R}(n,q,l+2^q) &= \{r : r = l + (x+1)2^q \text{ with } 0 \le x < 2^{2^n-q}\} \\
&= \{r : r = l + x2^q \text{ with } 1 \le x \le 2^{2^n-q}\} \\
&= \{r : r = l + x2^q \text{ with } 1 \le x < 2^{2^n-q}\} \cup \{l + 2^{2^n}\} \\
&= \{r : r = l + x2^q \text{ with } 0 \le x < 2^{2^n-q}\} \\
&\qquad\qquad\qquad\qquad\qquad\qquad \mod (F_n - 1)
\end{aligned}$$

$$\mathcal{S}(n,q,l+2^q) = \mathcal{S}(n,q,l)$$

$\square$

Lemma 76. *Let $0 \le q < 2^n - 1$. we may extend the definition*

(7.1.26) $\mathcal{M}(n,q,l) := \oplus \mathcal{S}(n,q+1,l) \times \mathcal{S}(n,q+1,l+2^q) =$

Multiset $\{3^r + 3^{r'} : r \in \mathcal{R}(n,q+1,l) \text{ and } r' \in \mathcal{R}(n,q+1,l+2^q)\}$

to all integer l. In that case holds

$$\mathcal{M}(n,q,l+2^q) = \mathcal{M}(n,q,l)$$

Proof. One may simplify in the same way as in the previous lemma 75 and go on

$$\begin{aligned}
\mathcal{S}(n,q,l+2^q) &= \mathcal{S}(n,q,l) \\
\mathcal{M}(n,q,l+2^q) &= \oplus \mathcal{S}(n,q+1,l+2^q) \times \mathcal{S}(n,q+1,l+2^{q+1}) \\
&= \oplus \mathcal{S}(n,q+1,l+2^q) \times \mathcal{S}(n,q+1,l) \\
&= \mathcal{M}(n,q,l)
\end{aligned}$$

$\square$

Lemma 77. *Let $0 \le q < 2^n - 1$ and $0 \le l < 2^q$. The above Fourier-typ analysis has to be used for the delicate multiset (7.1.26). Any two elements $3^s, 3^t \in \mathcal{M}(n,q,l)$ with $2^q \mid s - t$ occur with the same multiplicity. (They need not be equal.) Moreover there exist nonnegative coefficients $c_{[q,l,k]}$ for $0 \le l < 2^q$, $0 \le k < 2^q$ such that*

(7.1.27)
$$\oplus \mathcal{S}(n,q+1,l) \times \mathcal{S}(n,q+1,l+2^q) = \bigcup_{0 \le k < 2^q} c_{[q,l,k]} \mathcal{S}(n,q,k)$$

(7.1.28)
$$S(n,q+1,l) \cdot S(n,q+1,l+2^q) = \sum_{0 \le k < 2^q} c_{[q,l,k]} S(n,q,k)$$

7.1. FERMAT PRIMES

Remark. We can only prove that any two elements $3^s, 3^t \in \mathcal{M}(n,q,l)$ with $2^q \mid s-t$ occur with the same multiplicity, but they need not be equal.

Proof. Take any elements 3^s and 3^t with $2^q \mid s-t$ and assume 3^s occurs in the multiset $\mathcal{M}(n,q,l)$. By assumption $s = l + 2^q x$ and $t = l + 2^q y$ holds for some $0 \leq x, y < 2^{2^n - q}$ for which we may assume $y > x$. We may write

$$3^t = 3^{(l+(y-x)2^q)+2^q x} \in \mathcal{M}(n,q,l+(y-x)2^q) = \mathcal{M}(n,q,l)$$

to let a wonder happen. □

Lemma 78. *The coefficients from the last lemma 77 are*

$$2^{2^n - q} \cdot c_{[q,l,k]} = \#\mathcal{M}(n,q,l) \cap \mathcal{S}(n,q,k)$$
$$= \#\mathcal{S}(n,q,k) \cap [\oplus \mathcal{S}(n,q+1,l) \times \mathcal{S}(n,q+1,l+2^q)]$$

Hence $2^{2^n - q} \cdot c_{[q,l,k]}$ is the number of solutions (r, r', s) of the congruence

(7.1.29) $\qquad 3^r + 3^{r'} \equiv 3^s \pmod{F_n}$

with $r \in \mathcal{R}(n,q+1,l)$, $r' \in \mathcal{R}(n,q+1,l+2^q)$ and $s \in \mathcal{R}(n,q,k)$.

Definition 35. *Let $0 \leq q < 2^n$ and $a \equiv 3^{2^q} \pmod{F_n}$ and fix the number d. We define the coefficient $e[n,q,d]$ as the number of solutions of the congruence*

(7.1.30) $\quad a^{2x} + a^{1+2y} \equiv 3^d \pmod{F_n}$ with $0 \leq x, y < 2^{2^n - q - 1}$

We define the coefficient $ex0[n,q,d]$ as the number of solutions of the congruence

(7.1.31) $\quad 1 + a^{1+2y} \equiv 3^d \pmod{F_n}$ with $0 \leq y < 2^{2^n - q - 1}$

Lemma 79. *I go on to prepare the calculation of the coefficients $e[n,q,d]$.*

(i)

$$(7.1.32) \quad e[n,q,d] = \sum_{0 \leq x < 2^{2^n - q - 1}} ex0[n, q, d + x * 2^{q+1}]$$

(ii) *The coefficient $e[n,q,d]$ only depends on $d \mod 2^q$ and on $x \mod 2^{2^n - q - 1}$ and $y \mod 2^{2^n - q - 1}$*

(iii)

(7.1.33) $$\sum_{0\leq d<2^q} e[n,q,d] = 2^{2^n-q-2}$$

Proof. Fix n and define

$$t[x,y,q,d] := \left(3^{2x\cdot 2^q} + 3^{(1+2y)\cdot 2^q} - 3^d\right) \pmod{F_n}$$

Because of the Little Fermat Theorem, the quantity $t[x,y,q,d]$ depends only on $x \bmod 2^{2^n-q-1}$ and $y \bmod 2^{2^n-q-1}$ and $d \bmod 2^{2^n}$. The reader should check the identities

$$t[x,y,q,d]\cdot 3^{2^q} = t[y, x+1, d+2^q]$$
$$t[x,y,q,d]\cdot 3^{2^{q+1}} = t[x+1, y+1, d+2^{q+1}]$$
$$t[x,y,q,d]\cdot 3^{z2^{q+1}} = t[x+z, y+z, d+z2^{q+1}]$$

In terms of the characteristic function

$$\chi(P) = \begin{cases} 1 & \text{if P is true} \\ 0 & \text{if P is false} \end{cases}$$

equations (7.1.31) and (7.1.30) become

(7.1.34)
$$ex0[n,q,d] = \sum_{0\leq y<2^{2^n-q-1}} \chi(t[0,y,q,d]\equiv 0 \pmod{F_n})$$
$$e[n,q,d] = \sum_{0\leq x<2^{2^n-q-1}\text{ and }0\leq y<2^{2^n-q-1}} \chi(t[x,y,q,d]\equiv 0$$
$$\pmod{F_n})$$

In the following proof all congruences are understood modulo F_n.

(i) The sum from item (i) extends over the entire period. Hence substitutions $x \mapsto x+z$ and $x \mapsto -x$ leave them invariant.

7.1. FERMAT PRIMES

Finally, we use the identities for t and get

$$\sum_{0 \leq x < 2^{2^n-q-1}} ex0[n, q, d + x * 2^{q+1}]$$
$$= \sum_{xy} \chi(t[0, y, d + x * 2^{q+1}] \equiv 0)$$
$$= \sum_{xy} \chi(t[0, y - x, d - x * 2^{q+1}] \equiv 0)$$
$$= \sum_{xy} \chi(t[x, y, q, d] \equiv 0)$$
$$= e[n, q, d]$$

with $xy = 0 \leq x < 2^{2^n-q-1}$ and $0 \leq y < 2^{2^n-q-1}$ as claimed.

(ii) To confirm the shorter period 2^q for the variable d, we use the fact that the substitution $x, y \mapsto y, x + 1$ is allowed for the sums over the entire periods. Finally, we use the identities for t and get

$$e[n, q, d + 2^q] = \sum_{xy} \chi(t[x, y, d + 2^q] \equiv 0)$$
$$= \sum_{xy} \chi(t[y, x + 1, d + 2^q] \equiv 0)$$
$$= \sum_{xy} \chi(t[x, y, q, d] \equiv 0)$$
$$= e[n, q, d]$$

confirming period 2^q.

(iii) Since all these 3^d are inequivalent, for fixed x and y the first sum equals 1. Periodicity 2^q implies the last two lines

$$\sum_{0 \leq d < 2^{2^n}} \chi(t[x,y,q,d] \equiv 0) = 1$$

$$\sum_{0 \leq d < 2^{2^n} \text{ and } xy} \chi(t[x,y,q,d] \equiv 0) = 2^{2(2^n-q-1)}$$

$$\sum_{0 \leq d < 2^{2^n}} e[n,q,d] = 2^2(2^n - q - 1)$$

$$2^{2^n-q} \sum_{0 \leq d < 2^q} e[n,q,d] = 2^{2(2^n-q-1)}$$

$$\sum_{0 \leq d < 2^q} e[n,q,d] = 2^{2^n-q-2}$$

$\square$

I go on with some counting.

Lemma 80. *For $0 \leq q < 2^n - 1$ and $0 \leq d < 2^q$, the counting (7.1.33) has the following consequences:*

(a) *If for some fixed value q all $e[n,q,d]$ are equal, then $e[n,q,d] = 2^{2^n-2q-2}$. Especially we get $e[n,0,0] = 2^{2^n-2}$. But this simple outcome is only possible for $q \leq 2^{n-1} - 1$.*

(b) *For the range $2^{n-1} \leq q \leq 2^n - 2$ at most 2^{2^n-q-2} among the coefficients $e[n,q,d]$ are nonzero. Note that*

$$2^q > 2^{2^{n-1}-2} \geq 2^{2^n-q-2} \geq 1$$

If the number of coefficients $e[n,q,d] \neq 0$ is maximal, then all these coefficients are zero or one.

Especially for $q = 2^n - 2$, there exists a unique index $d*$ such that
$$e[n, 2^n - 2, d] = \delta_{d,d*}$$

Lemma 81. *Let $0 \leq q < 2^n$*

(i) *Let $0 \leq l, k < 2^q$ and $d \equiv (k - l) \pmod{2^q}$. The number of solutions (r, r', s) of the congruence*

(7.1.29) $\quad 3^r + 3^{r'} \equiv 3^s \pmod{F_n}$

with $r \in \mathcal{R}(n, q+1, l)$, $r' \in \mathcal{R}(n, q+1, l+2^q)$ and $s \in \mathcal{R}(n, q, k)$ equals $2^{2^n-q} \cdot e[n,q,d]$.

7.1. FERMAT PRIMES

(ii) $c_{[q,l,k]} = e[n,q,d]$ *holds for* $d \equiv (k-l) \mod 2^q$. *Especially the coefficient* $e[n,q,d]$ *depends only on* $d \mod 2^q$.
Formula (7.1.27) *simplifies to*

$$\text{(7.1.35)} \quad \oplus \mathcal{S}(n,q+1,l) \times \mathcal{S}(n,q+1,l+2^q)$$
$$= \bigcup_{0 \leq d < 2^q} e[n,q,d]\mathcal{S}(n,q,l+d)$$

$$\text{(7.1.36)} \quad S(n,q+1,l) \cdot S(n,q+1,l+2^q)$$
$$= \sum_{0 \leq d < 2^q} e[n,q,d]S(n,q,l+d)$$

for $0 \leq l < 2^q$.

Proof. (i) Let $0 \leq l, k < 2^q$ and $d \equiv (k-l) \mod 2^q$. The number of solutions (r, r', s) of the congruence

$$\text{(7.1.29)} \quad 3^r + 3^{r'} \equiv 3^s \pmod{F_n}$$

has to be counted under the restrictions $r \in \mathcal{R}(n, q+1, l)$, $r' \in \mathcal{R}(n, q+1, l+2^q)$ and $s \in \mathcal{R}(n, q, k)$. These restriction are

$$r = l + (2x)2^q \text{ with } 0 \leq 2x < 2^{2^n-q}$$
$$r' = l + (1+2y)2^q \text{ with } 0 \leq 2y < 2^{2^n-q}$$
$$s \equiv l + d + z2^q \pmod{2^q} \text{ with } 0 \leq z < 2^{2^n-q}$$

and the congruence (7.1.29) is equivalent to

$$3^l \cdot a^{2x} + 3^l \cdot a^{1+2y} \equiv 3^l \cdot 3^{d+\epsilon 2^q} \cdot a^{z-\epsilon} \pmod{F_n}$$

Here
$$\epsilon = \begin{cases} 0 & \text{if } d \geq 0; \\ 1 & \text{if } d < 0. \end{cases}$$

We need to simplify separately in four cases.

- Assume $d \geq 0$ and $z = 2u$. One gets

$$a^{2(x-u)} + a^{1+2(y-u)} \equiv 3^d \pmod{F_n}$$

- Assume $d \geq 0$ and $z = -1 + 2u$. One gets

$$a^{1+2(x-u)} + a^{2(y-u+1)} \equiv 3^d \pmod{F_n}$$

- Assume $d < 0$ and $z = 2u$. One getts

$$a^{1+2(x-u)} + a^{2(y-u+1)} \equiv 3^{d+2^q} \pmod{F_n}$$

- Assume $d < 0$ and $z = 1 + 2u$. One gets

$$a^{2(x-u)} + a^{1+2(y-u)} \equiv 3^{d+2^q} \pmod{F_n}$$

In each one of the four cases, the number of solutions for fixed z is equal to $e[n,q,d]$. Since z may assume 2^{2^n-q} values, counting both odd and even cases, the total number of solutions of equation (7.1.29) is $2^{2^n-q} \cdot e[n,q,d]$.

(ii) This item follows by using item (i) and the last lemma 78. □

Corollary 30 (Gauss' main result). *For all Fermat primes, the regular F_n-gon is constructible by ruler and compass.*

Remark. The first explicit constructions of a regular 257-gon were given by Magnus Georg Paucker (1822) [7] and Friedrich Julius Richelot (1832). A construction for a regular 65537-gon was first given by Johann Gustav Hermes (1894) [5]. The construction is very complex; Hermes spent 10 years completing the 200-page manuscript.

Proof. We just recapitulate the procedure outlined above. The regular F_n-gon is put into the complex plane with all vertices on the unit circle, and one vertex at $z = 1$. The remaining vertices are obtained from the formula

$$(7.1.8) \qquad d_r := \exp\frac{2\pi i \cdot 3^r}{F_n} \quad \text{with } 0 \le r < F_n - 1.$$

At each vertex of the rooted binary tree with 2^n levels of descendents is put the set

$$(7.1.12) \qquad \mathcal{R}(n,q,l) = \{r : r = l + x2^q \text{ with } 0 \le x < 2^{2^n-q}\}$$

At first are calculated the sums of vertices

$$(7.1.14) \qquad S(n,q,l) = \sum \{\exp\frac{2\pi i \cdot k}{F_n} : k \in \mathcal{S}(n,q,l)\}$$
$$= \sum \{\exp\frac{2\pi i \cdot 3^r}{F_n} : r \in \mathcal{R}(n,q,l)\}$$

7.1. FERMAT PRIMES

This binary tree has a root which gets the level $q = 0$. At the root is put the sum $s = -1$, which is the sum of all d_r. The corresponding binary integers are the set

(7.1.10) $\quad \mathcal{R}(n, 0, 0) = \{ \sum_{0 \leq q < 2^n} b_q 2^q : b_q \in \{0, 1\} \text{ for } 0 \leq q < 2^n \}$

Recursively are computated the sums $S(n, q, l)$ at the 2^q vertices of the q-th level. Here $0 \leq q \leq 2^n$ is the level in the tree and $0 \leq l < 2^q$ enumerates the vertices in a given level, say from left to right. Assume that the sums $S(n, q, l)$ for the q-th level are already known. For the next level, the sums $S(n, q+1, l)$ and $S(n, q+1, l+2^q)$ are the roots of the quadratic equation $x^2 - Px + Q = 0$ with the coefficients

$P = S(n, q+1, l) + S(n, q+1, l+2^q) = S(n, q, l)$

(7.1.28)

$Q = S(n, q+1, l) \cdot S(n, q+1, l+2^q) = \sum_{0 \leq d < 2^q} e[n, q, d] S(n, q, l+d)$

The coefficients P and Q indeed depend only on the sums of the q-th level and are already recursively known. Indeed, by lemma 81 the coefficient $e[n, q, d]$ for $0 \leq q < 2^n - 1$ and $0 \leq d < 2^q$ is the number of solutions of the congruence

(7.1.30) $\qquad a^{2x} + a^{1+2y} \equiv 3^d \pmod{F_n}$

with

$a \equiv 3^{2^q} \pmod{F_n}$ and $0 \leq x, y < 2^{2^n - q - 1}$

From the well-known formula for the roots of a quadratic equation, we obtain

(7.1.37) $\quad \Delta(n, q, l) = S(n, q, l)^2 - 4 \sum_{0 \leq d < 2^q} e[n, q, d] S(n, q, l+d)$

(7.1.38)
$$S(n, q+1, l) = \frac{S(n, q, l) + \sigma(n, q, l) \sqrt{\Delta(n, q, l)}}{2}$$

(7.1.39)
$$S(n, q+1, l+2^q) = \frac{S(n, q, l) - \sigma(n, q, l) \sqrt{\Delta(n, q, l)}}{2}$$

with the sign $\sigma(n, q, l) = \pm 1$, which has to be fixed numerically. For all levels $q < 2^n - 1$, all quantities are real. One obtains the result

(7.1.40) $\qquad 2\cos\dfrac{2\pi \cdot 3^l}{F_n} = S(n, 2^n - 1, l) \quad \text{for all } 0 \leq l < h$

with

(7.1.16) $\qquad h = \dfrac{F_n - 1}{2} = 2^{2^n - 1}$

As explained by lemma 72, the last level with $q = 2^n - 1$ yields the complex solutions

$$S(n, 2^n, l), \; S(n, 2^n, l + h) = \exp \pm \dfrac{2\pi i \cdot 3^l}{F_n}$$

$\square$

Remark. Theoretically, the corollary 30 solves the construction problem <u>for all</u> possible F_n-gons, even in the case that some huge still unknown Fermat prime should exist.

On the other hand, so many calculations are involved that the result is totally impractical. One needs to work towards further simplifications to go beyond the 17-gon.

With the next corollary 31 I skip ahead a bid and use the numerical results explained below. Especially, this means that corollary 31 and remark 7.1.3 only refer to the <u>known</u> Fermat primes, but not to any further hugh Fermat prime in case it might exist.

Corollary 31. *After having done numerical calculations for $F_2 = 17$, $F_3 = 257$, and $F_4 = 65\,537$, one gets the following results confirming the above lemma:*

(a) *For $q = 0$ and $q = 1$, all coefficients $e[n, q, d]$ are equal. Hence we get $e[n, 0, 0] = 2^{2^n - 2}$ and $e[n, 1, 0] = e[n, 1, 1] = 2^{2^n - 4}$. This simple outcome does not occur for any $q \geq 2$.*

(b) *For $n = 2, q \geq 2$ and $n = 3, q \geq 4$ and $n = 4, q \geq 11$, the number of coefficients $e[n, q, d] \neq 0$ is maximal, and hence all these coefficients are equal to one.*

Especially for $q = 2^n - 2$, there exists a unique index $d*$ such that
$e[n, 2^n - 2, d] = \delta_{d, d*}$. One gets $d = 1$ for $n = 2$, and $d* = 56$ for $n = 3$, and $d* = 3072$ for $n = 4$.*

7.1. FERMAT PRIMES

(c) For $n = 3, q = 2, 3$ and $n = 4, 2 \leq q \leq 10$, the coefficients $e[n, q, d]$ take much more random values than I originally did expect. I cannot see any general simple pattern.

Remark. With the values from Corollary 31 one gets for $q = 0$ as already obtained in lemma 73

$$Q(n, 0, 0) = e[n, 0, 0]S(n, 0, 0) = -2^{2^n - 2}$$
$$\Delta(n, 0, 0) = S(n, 0, 0)^2 - 4Q(n, 0, 0) = F_n$$
$$S(n, 1, 0) = \frac{-1 + \sigma(n, 0, 0)\sqrt{F_n}}{2}$$
$$S(n, 1, 1) = \frac{-1 - \sigma(n, 0, 0)\sqrt{F_n}}{2}$$

For $q = 1$ we obtain for $n = 2, 3, 4$

$$4S(2, 2, 0) = -1 + \sqrt{17} + \sqrt{34 - 2\sqrt{17}}$$
$$4S(3, 2, 0) = -1 + \sqrt{257} + \sqrt{514 - 2\sqrt{257}}$$
$$4S(4, 2, 0) = -1 + \sqrt{65\,537} - \sqrt{131\,074 - 2\sqrt{65\,537}}$$

The first value agrees with the result obtained earlier since

$$4S(2, 2, 0) = 4c_1 + 4c_4 = 8\cos\frac{2\pi}{17} + 8\cos\frac{8\pi}{17}$$
$$= -1 + \sqrt{17} + \sqrt{34 - 2\sqrt{17}}$$

Problem 98. Assume that $e[n, 1, 0] = e[n, 1, 1] = 2^{2^n - 4}$ and $\sigma[n, 0, 0] = 1$ hold ,—not only for $n = 2, 3, 4$ as we shall check numerically,—but for all n for which F_n turns out to be a Fermat prime. Check the resulting formulas

$$S(n, 2, 0) = \frac{-1 + \sqrt{F_n} + \sigma(n, 1, 0)\sqrt{2F_n - 2\sqrt{F_n}}}{4}$$
$$S(n, 2, 2) = \frac{-1 + \sqrt{F_n} - \sigma(n, 1, 0)\sqrt{2F_n - 2\sqrt{F_n}}}{4}$$
$$S(n, 2, 1) = \frac{-1 - \sqrt{F_n} + \sigma(n, 1, 1)\sqrt{2F_n + 2\sqrt{F_n}}}{4}$$
$$S(n, 2, 2) = \frac{-1 - \sqrt{F_n} - \sigma(n, 1, 1)\sqrt{2F_n + 2\sqrt{F_n}}}{4}$$

Solution.

$$S(n,1,0) = \frac{-1+\sqrt{F_n}}{2} \quad \text{and} \quad S(n,1,1) = \frac{-1-\sqrt{F_n}}{2}$$

$$\begin{aligned}
Q(n,1,l) &= e[n,1,0]S(n,1,l) + e[n,1,1]S(n,1,l+1) \\
&= 2^{2^n-4}(S(n,1,l) + S(n,1,l+1)) = -2^{2^n-4} \\
\Delta(n,1,0) &= S(n,1,0)^2 - 4Q(n,1,0) \\
&= \frac{1+F_n - 2\sigma(n,0,0)\sqrt{F_n}}{4} + 2^{2^n-2} = \frac{F_n - \sqrt{F_n}}{2} \\
\Delta(n,1,1) &= \frac{F_n + \sqrt{F_n}}{2}
\end{aligned}$$

$$\begin{aligned}
S(n,2,0) &= \frac{S(n,1,0) + \sigma(n,1,0)\sqrt{\Delta(n,1,0)}}{2} \\
&= \frac{-1+\sqrt{F_n} + 2\sigma(n,1,0)\sqrt{\Delta(n,1,0)}}{4} \\
&= \frac{-1+\sqrt{F_n} + \sigma(n,1,0)\sqrt{2F_n - 2\sqrt{F_n}}}{4}
\end{aligned}$$

$$S(n,2,2) = \frac{-1+\sqrt{F_n} - \sigma(n,1,0)\sqrt{2F_n - 2\sqrt{F_n}}}{4}$$

$$S(n,2,1) = \frac{-1-\sqrt{F_n} + \sigma(n,1,1)\sqrt{2F_n + 2\sqrt{F_n}}}{4}$$

$$S(n,2,2) = \frac{-1-\sqrt{F_n} - \sigma(n,1,1)\sqrt{2F_n + 2\sqrt{F_n}}}{4}$$

□

These beginnings are still too simple to guess how to go on. Here are some steps I did to gather confidence for the heavy numerics to follow. Next I try to get more information from my pairing method earlier used successfully for the 17-gon. Take the next example with $F_n = 257$, thus $n = 3$ and $h = 128$. Let $\alpha = \frac{2\pi}{257}$. For the level $q = 2^n - 2 = 6$ occur the 64 pairs of double cosines $2\cos k\alpha$. One of them is

$$3^0,\ 3^{64},\ 3^{128},\ 3^{192} \equiv 1,\ -16,\ -1,\ 16 \pmod{257}$$

7.1. FERMAT PRIMES 203

Further ones are obtained from the addition theorem (7.1.2) of the cosin function to be

$$\{1, 16\}, \{15, 17\}, \{2, 32\}, \{30, 34\},$$
$$\{4, 64\}, \{60, 68\}, \{8, 128\}, \{120, 136\}$$

The next pair $\{16, 256\} \equiv \{1, 16\}$ closes the ring. Hence there exist 8 such rings of each 8 cosine pairs.

Translate into Gauss' approach with the primitive root.

Remark. The pairs $(3^r \mod 257.r)$ are

((1 . 0) (2 . 48) (3 . 1) (4 . 96) (5 . 55) (6 . 49) (7 . 85) (8 . 144) (9 . 2) (10 . 103) (11 . 196) (12 . 97) (13 . 106) (14 . 133) (15 . 56) (16 . 192) (17 . 120) (18 . 50) (19 . 125) (20 . 151) (21 . 86) (22 . 244) (23 . 28) (24 . 145) (25 . 110) (26 . 154) (27 . 3) (28 . 181) (29 . 94) (30 . 104) (31 . 242) (32 . 240) (33 . 197) (34 . 168) (35 . 140) (36 . 98) (37 . 219) (38 . 173) (39 . 107) (40 . 199) (41 . 19) (42 . 134) (43 . 207) (44 . 36) (45 . 57) (46 . 76) (47 . 61) (48 . 193) (49 . 170) (50 . 158) (51 . 121) (52 . 202) (53 . 89) (54 . 51) (55 . 251) (56 . 229) (57 . 126) (58 . 142) (59 . 118) (60 . 152) (61 . 138) (62 . 34) (63 . 87) (64 . 32) (65 . 161) (66 . 245) (67 . 100) (68 . 216) (69 . 29) (70 . 188) (71 . 163) (72 . 146) (73 . 44) (74 . 11) (75 . 111) (76 . 221) (77 . 25) (78 . 155) (79 . 22) (80 . 247) (81 . 4) (82 . 67) (83 . 15) (84 . 182) (85 . 175) (86 . 255) (87 . 95) (88 . 84) (89 . 102) (90 . 105) (91 . 191) (92 . 124) (93 . 243) (94 . 109) (95 . 180) (96 . 241) (97 . 167) (98 . 218) (99 . 198) (100 . 206) (101 . 75) (102 . 169) (103 . 201) (104 . 250) (105 . 141) (106 . 137) (107 . 31) (108 . 99) (109 . 187) (110 . 43) (111 . 220) (112 . 21) (113 . 66) (114 . 174) (115 . 83) (116 . 190) (117 . 108) (118 . 166) (119 . 205) (120 . 200) (121 . 136) (122 . 186) (123 . 20) (124 . 82) (125 . 165) (126 . 135) (127 . 81) (128 . 80) (129 . 208) (130 . 209) (131 . 7) (132 . 37) (133 . 210) (134 . 148) (135 . 58) (136 . 8) (137 . 72) (138 . 77) (139 . 38) (140 . 236) (141 . 62) (142 . 211) (143 . 46) (144 . 194) (145 . 149) (146 . 92) (147 . 171) (148 . 59) (149 . 227) (150 . 159) (151 . 9) (152 . 13) (153 . 122) (154 . 73) (155 . 41) (156 . 203) (157 . 78) (158 . 70) (159 . 90) (160 . 39) (161 . 113) (162 . 52) (163 . 237) (164 . 115) (165 . 252) (166 . 63) (167 . 233) (168 . 230) (169 . 212) (170 . 223) (171 . 127) (172 . 47) (173 . 54) (174 . 143) (175 . 195) (176 . 132) (177 . 119) (178 . 150) (179 . 27) (180 . 153) (181 . 93) (182 . 239) (183 . 139) (184 . 172) (185 . 18) (186 . 35) (187 . 60) (188 . 157) (189 . 88) (190 . 228) (191 . 117) (192 . 33) (193 . 160) (194 . 215) (195 . 162) (196 . 10) (197 . 24) (198

. 246) (199 . 14) (200 . 254) (201 . 101) (202 . 123) (203 . 179) (204 . 217) (205 . 74) (206 . 249) (207 . 30) (208 . 42) (209 . 65) (210 . 189) (211 . 204) (212 . 185) (213 . 164) (214 . 79) (215 . 6) (216 . 147) (217 . 71) (218 . 235) (219 . 45) (220 . 91) (221 . 226) (222 . 12) (223 . 40) (224 . 69) (225 . 112) (226 . 114) (227 . 232) (228 . 222) (229 . 53) (230 . 131) (231 . 26) (232 . 238) (233 . 17) (234 . 156) (235 . 116) (236 . 214) (237 . 23) (238 . 253) (239 . 178) (240 . 248) (241 . 64) (242 . 184) (243 . 5) (244 . 234) (245 . 225) (246 . 68) (247 . 231) (248 . 130) (249 . 16) (250 . 213) (251 . 177) (252 . 183) (253 . 224) (254 . 129) (255 . 176) (256 . 128))

$$3^0,\ 3^{64},\ 3^{128},\ 3^{192} \equiv 1,\ -16,\ -1,\ 16 \pmod{257}$$
$$3^{56},\ 3^{120},\ 3^{184},\ 3^{248} \equiv 15,\ 17,\ -15,\ -17 \pmod{257}$$
$$3^{48},\ 3^{112},\ 3^{176},\ 3^{240} \equiv 2,\ -32,\ -2,\ 32 \pmod{257}$$
$$3^{40},\ 3^{104},\ 3^{168},\ 3^{232} \equiv -34,\ 30,\ 34,\ -30 \pmod{257}$$
$$3^{32},\ 3^{96},\ 3^{160},\ 3^{224} \equiv 64,\ 4,\ -64,\ -4 \pmod{257}$$
$$3^{24},\ 3^{88},\ 3^{152},\ 3^{216} \equiv -60,\ -68,\ 60,\ 68 \pmod{257}$$
$$3^{16},\ 3^{80},\ 3^{144},\ 3^{208} \equiv -8,\ 128,\ 8,\ -128 \pmod{257}$$
$$3^8,\ 3^{72},\ 3^{136},\ 3^{200} \equiv 136,\ -120,\ -136,,\ 120 \pmod{257}$$
$$3^0,\ 3^{64},\ 3^{128},\ 3^{192} \equiv 1,\ -16,\ -1,\ 16 \pmod{257}$$

We may now determine $d*$ in lemma 80. As

$$S(3,6,0) = \{3^0,\ 3^{64},\ 3^{128},\ 3^{192} \mod 257\} = \{1,\ -16,\ -1,\ 16\}$$

is subdivided into the two pairs $S(3,7,0) = \{\pm 1\}$ and $S(3,7,2^6) = \{\pm 16\}$ we need (for the setup of the quadratic equations) to get

$$\oplus S(3,7,0) \times S(3,7,2^6) = \oplus\{1,-1\} \times \{-16,16\} = \{\pm 15, \pm 17\}$$
$$= \{3^{56},\ 3^{120},\ 3^{184},\ 3^{248} \mod 257\} = S(3,6,56)$$

Hence $d* = 56$ in lemma 80.

We go back one level to $q = 5$. There occur 32 groups of 8. We take the one with $l = 0$.

$$S(3,5,0) = \{3^0,\ 3^{32},\ 3^{64},\ 3^{96},\ 3^{128},\ 3^{160},\ 3^{192},\ 3^{224} \mod 257\}$$
$$= \{1,\ 64,\ -16,\ 4,\ -1,\ -64,\ 16,\ -4\}$$

It is subdivided into

$$S(3,6,0) = \{\pm 1, \pm 16\} \text{ and } S(3,6,2^5) = \{\pm 4, \pm 64\}$$

7.1. FERMAT PRIMES

We have to realize the formula (7.1.35), it is enough with say $l = 0$,

$$\mathcal{M}(n,q) := \oplus \mathcal{S}(n, q+1, 0) \times \mathcal{S}(n, q+1, 2^q)$$
$$= \bigcup_{0 \leq d < 2^q} e[n, q, d]\mathcal{S}(n, q, d)$$
$$\oplus \mathcal{S}(3, 6, 0) \times \mathcal{S}(3, 6, 2^5) = \bigcup_{0 \leq d < 32} e[3, 5, d]\mathcal{S}(3, 5, d)$$
$$\log_3 \mathcal{M}(n, q) = \bigcup_{0 \leq d < 2^q} e[3, q, d]\mathcal{R}(n, q, d)$$
$$\frac{\log_3 \mathcal{M}(n, q) \cap \mathcal{R}(n, q, d)}{2^{2n-q}} = e[3, q, d]$$

and extract from the present instance the coefficients $e[3, 5, d]$.

$$\oplus \mathcal{S}(3, 6, 0) \times \mathcal{S}(3, 6, 2^5)$$
$$= \oplus\{\pm 1, \pm 16\} \times \{\pm 4, \pm 64\}$$
$$= \{\pm 3, \pm 5, \pm 63, \pm 65, \pm 12, \pm 20, \pm 48, \pm 80 \mod 257\}$$
$$= 3^{\{1,55,87,161,97,151,193,247\}(+128)}$$
$$= 3^{\{1,129,55,183,87,215,161,33,97,225,151,23,193,65,247,119\}}$$
$$= 3^{\{23,33,55,65,87,97,119,129,151,161,183,193,215,225,247,257\}}$$
$$= 3^{\{23,55,87,119,151,183,215,247\} \cup 10+\{23,55,87,119,151,183,215,247\}}$$
$$= S(3, 5, 23) \cup S(3, 5, 33)$$

Hence $e[3, 5, 23] = e[3, 5, 33] = 1$ and all other $e[3, 5, d] = 0$.

Remark. By lemma 81 item(iii) the coefficient $e[n, q, d]$ depends only on $d \mod 2^q$. Hence we obtain $= e[3, 5, 1] = e[3, 5, 23] = 1$ which agrees with the numerical result of proposition 62 below.

For the level $q = 4$ there occur 16 groups of 16. One of them is

$$3^0, 3^{16}, 3^{32}, 3^{48}, 3^{64}, 3^{80}, 3^{96}, 3^{112}, 3^{128}, 3^{144}, 3^{160}, 3^{176},$$
$$3^{192}, 3^{208}, 3^{224}, 3^{240}$$
$$\equiv 1, -8, 64, 2, -16, 128, 4, -32, -1, 8, -64, -2,$$
$$16, -128, -4, 32 \pmod{257}$$

For the level $q = 3$ there occur 8 groups of 32. One of them is

$$3^0, \, 3^8, \, 3^{16}, \, 3^{24}, \, 3^{32}, \, 3^{40}, \, 3^{48}, \, 3^{56}, \, 3^{64}, \, 3^{72}, \, 3^{80}, \, 3^{88},$$
$$3^{96}, \, 3^{104}, \, 3^{112}, \, 3^{120},$$
$$3^{128}, \, 3^{136},, \, 3^{144}, \, 3^{152}, \, 3^{160}, \, 3^{168}, \, 3^{176}, \, 3^{184}, \, 3^{192}, \, 3^{200},$$
$$3^{208}, \, 3^{216}, \, 3^{224}, \, 3^{232}, \, 3^{240}, \, 3^{248}$$
$$\equiv 1, \, -121, \, -8, \, -60, \, 64, \, 34, \, 2, \, 15, \, -16, \, 137, \, 128, \, -68,$$
$$4, \, 30, \, -32, \, 17,$$
$$-1, \, 121, \, 8, \, 60, \, -64, \, 34, \, -2, \, -15, \, 16, \, 120,$$
$$-128, \, 68, \, -4, \, -30, \, 32, \, -17 \pmod{257}$$

I have made further guesses for the 257-gon, but they continued to turn out wrong. Finally, it doomed to me that there is no easy to guess pattern for the neither the 257-gon nor the 65 537-gon. Here is the computer program and the result obtained with the help of DrRacket:

```
(require math/base)
(require math/number-theory)

(define (Fermat n)(add1(expt 2(expt 2 n))))
(define proot 3)

;a^{2x} + a^{1+2y} \equiv 3^{d} \pmod{F_n}
;a \equiv   3^{2^q} \pmod {F_n} \,\text{ and } \;
;0\le x,y  < 2^{2^n-q-1}%\,,\; 0\le y < 2^{2^n-q-1}

(define (eshort n q)
  (define 2q (expt  2 q))
  (define large(expt 2(sub1(-(expt 2 n)q))))
  (define a(with-modulus(Fermat n)
                  (modexpt proot (expt 2 q))))
  (define (iter x y d incid result)
  (cond [(equal? d 2q) result]
        [(equal? y large)
           (if(equal? incid 0)
              (iter 0 0 (add1 d) 0 result)
              (iter 0 0 (add1 d) 0
                 (cons(cons d incid)result)))]
        [(equal? x large)
            (iter 0 (add1 y) d incid result)]
```

7.1. FERMAT PRIMES

```
        [else (let*[(axy(with-modulus(Fermat n)
                   (mod+(modexpt a(* 2 x))
                        (modexpt a(add1(* 2 y))))))
             (3d (with-modulus(Fermat n) (modexpt proot d)))
                (try(with-modulus(Fermat n)(mod- axy 3d)))]
                 (if (equal? try 0)
                    (iter(add1 x)y d (add1 incid) result)
                    (iter(add1 x)y d incid result)))]))
   ; body of eshort n q calls iter:
    (reverse (iter 0 0 0 0 null)))

(define (R n q low)
   (let [(r(lambda(x)
        (with-modulus(sub1(Fermat n))
          (mod+ low(mod* x(modexpt 2 q))))))
         (large(expt 2(- (expt 2 n)q)))]
            (build-list large (lambda(i)(r i)))))

(define (S n q l)
   (map (lambda(r)
         (with-modulus (Fermat n) (modexpt proot r)))
        (R n q l)))

(define (Sum n q l)
   (let*[(alpha(/(* 2 pi)(Fermat n)))
         (lmod (with-modulus(expt 2 q) l))
         (anglelist(map(lambda(k)
              (\cos(* k alpha)))(S n q lmod)))]
       (foldl + 0 anglelist)))

(define (Q n q l)
  (define Qterms (map(lambda(eshortitem)
        (let*[(d(car eshortitem))
              (multi(cdr eshortitem))]
          (* multi(Sum n q (+ 1 d)))))
     (eshort n q)))
   (foldl + 0 Qterms))

(define (sigma n q l)
 (let*[(S (Sum n q l))
       (Delta (-(sqr S)(* 4 (Q n q l))))
       (top(-(* 2 (Sum n (add1 q)l))S))]
```

```
            (inexact->exact(round(/ top(sqrt Delta)))) ))

(displayln (let[(n 3)]
   (map (lambda(q)
        (let* [(enq(eshort n q))]
        (for-each (lambda(l)
          (let[(item (list-ref enq l))]
          (printf " e[~a,~a,~a]=" n q (car item))
          (display (cdr item )))
          (printf ",") null)
          (build-list(length enq) values)))
      (printf "\n")"end")
      (build-list(-(expt 2 n)1)values))))

(displayln (let[(n 3)]
   (map (lambda(q)
     (for-each (lambda(l)
        (printf " sigma(~a,~a,~a)="  n q l)
          (display (sigma n q l))null)
   (build-list(expt 2 q)values)) (displayln "")"end")
   (build-list(-(expt 2 n)1)values))))

(displayln (let[(n 4)]
   (map (lambda(q)
        (let* [(enq(eshort n q))]
        (for-each (lambda(l)
            (let[(item (list-ref enq l))]
            (printf " e[~a,~a,~a]=" n q (car item))
            (display (cdr item )))
            (printf ",") null)
          (build-list(length enq) values)))
      (printf "\n")"end")
      (build-list(-(expt 2 n)1)values))))

(displayln (let[(n 4)]
   (map (lambda(q)
     (for-each (lambda(l)
        (printf " sigma(~a,~a,~a)="  n q l)
          (display (sigma n q l))null)
   (build-list(expt 2 q)values)) (displayln "")"end")
   (build-list(-(expt 2 n)1)values))))
```

7.1. FERMAT PRIMES

Proposition 61. *Let $n = 2$ and $F_2 = 17$. As always, with primitive root 3, one obtains the nonzero structural constants $e[2, q, d] \neq 0$ with $q = 0, 1, 2$ and $0 \leq d < 2^q$ and the signs $\sigma(n, q, l)$ to be*

$$e[2, 0, 0] = 4,$$
$$e[2, 1, 0] = 1, e[2, 1, 1] = 1,$$
$$e[2, 2, 1] = 1,$$
$$\sigma(2, 0, 0) = 1,$$
$$\sigma(2, 1, 0) = 1, \ \sigma(2, 1, 1) = 1,$$
$$\sigma(2, 2, 0) = 1, \ \sigma(2, 2, 1) = 1, \ \sigma(2, 2, 2) = -1, \ \sigma(2, 2, 3) = -1$$

Proposition 62. *Let $n = 3$ and $F_3 = 257$. The nonzero structural constants $e[3, q, d] \neq 0$ with $q = 0, 1, 2, 3, 4, 5, 6$ and $0 \leq d < 2^q$ are the following ones*

$$e[3, 0, 0] = 64, \ e[3, 1, 0] = e[3, 1, 1] = 16,$$
$$e[3, 2, 0] = 2, \ e[3, 2, 1] = 5, \ e[3, 2, 2] = 4, \ e[3, 2, 3] = 5,$$
$$e[3, 3, 0] = 2, \ e[3, 3, 2] = 2, \ e[3, 3, 4] = 1, \ e[3, 3, 5] = 2,$$
$$e[3, 3, 6] = 1, \ e[3, 4, 0] = e[3, 4, 1] = e[3, 4, 2] = e[3, 4, 5] = 1,$$
$$e[3, 5, 1] = e[3, 5, 23] = 1, \ e[3, 6, 56] = 1$$

Remark. Many of the coefficients $e[3, q, d]$ depend on the choice of the primitive root a, as is shown by the following.

$$proot = 3$$
$$e[3, 0, 0] = 64, e[3, 1, 0] = 16, e[3, 1, 1] = 16,$$
$$e[3, 2, 0] = 2, e[3, 2, 1] = 5, e[3, 2, 2] = 4, e[3, 2, 3] = 5,$$
$$e[3, 3, 0] = 2, e[3, 3, 2] = 2, e[3, 3, 4] = 1, e[3, 3, 5] = 2, e[3, 3, 6] = 1,$$
$$e[3, 4, 0] = 1, e[3, 4, 1] = 1, e[3, 4, 2] = 1, e[3, 4, 5] = 1,$$
$$e[3, 5, 1] = 1, e[3, 5, 23] = 1, e[3, 6, 56] = 1$$

$$proot = 5$$
$$e[3, 0, 0] = 64, e[3, 1, 0] = 16, e[3, 1, 1] = 16,$$
$$e[3, 2, 0] = 2, e[3, 2, 1] = 5, e[3, 2, 2] = 4, e[3, 2, 3] = 5,$$
$$e[3, 3, 0] = 2, e[3, 3, 2] = 1, e[3, 3, 3] = 2, e[3, 3, 4] = 1,$$
$$e[3, 3, 6] = 2, e[3, 4, 0] = 1, e[3, 4, 3] = 1, e[3, 4, 7] = 1,$$
$$e[3, 4, 14] = 1, e[3, 5, 1] = 1, e[3, 5, 7] = 1, e[3, 6, 8] = 1$$

$$proot = 7$$
$$e[3,0,0] = 64, e[3,1,0] = 16, e[3,1,1] = 16,$$
$$e[3,2,0] = 2, e[3,2,1] = 5, e[3,2,2] = 4, e[3,2,3] = 5,$$
$$e[3,3,0] = 2, e[3,3,1] = 2, e[3,3,2] = 2, e[3,3,4] = 1, e[3,3,6] = 1,$$
$$e[3,4,0] = 1, e[3,4,1] = 1, e[3,4,10] = 1, e[3,4,13] = 1,$$
$$e[3,5,27] = 1, e[3,5,29] = 1, e[3,6,24] = 1$$

$$proot = 10$$
$$e[3,0,0] = 64, e[3,1,0] = 16, e[3,1,1] = 16,$$
$$e[3,2,0] = 2, e[3,2,1] = 5, e[3,2,2] = 4, e[3,2,3] = 5,$$
$$e[3,3,0] = 2, e[3,3,2] = 2, e[3,3,4] = 1, e[3,3,5] = 2, e[3,3,6] = 1,$$
$$e[3,4,0] = 1, e[3,4,1] = 1, e[3,4,2] = 1, e[3,4,5] = 1,$$
$$e[3,5,7] = 1, e[3,5,17] = 1, e[3,6,56] = 1$$

$$proot = 14$$
$$e[3,0,0] = 64, e[3,1,0] = 16, e[3,1,1] = 16,$$
$$e[3,2,0] = 2, e[3,2,1] = 5, e[3,2,2] = 4, e[3,2,3] = 5,$$
$$e[3,3,0] = 2, e[3,3,1] = 2, e[3,3,2] = 2, e[3,3,4] = 1, e[3,3,6] = 1,$$
$$e[3,4,0] = 1, e[3,4,1] = 1, e[3,4,10] = 1, e[3,4,13] = 1,$$
$$e[3,5,11] = 1, e[3,5,13] = 1, e[3,6,24] = 1,$$

Remark. We may now check the entire procedure. It is straightforward to calculate the sums $S(n,q,l)$ numerically. We solve equations (7.1.37) and (7.1.38) for the sign $\sigma(n,q,l)$.

$$Q(n,q,l) = \sum_{0 \leq d < 2^q} e[n,q,d]S(n,q,l+d)$$
$$\Delta(n,q,l) = S(n,q,l)^2 - 4Q(n,q,l)$$
$$\sigma(n,q,l) = \frac{2S(n,q+1,l) - S(n,q,l)}{\sqrt{\Delta(n,q,l)}}$$

These signs should turn out to be approximately ± 1, with small numerical defects. Indeed I have done this procedure successfully for $F_2 = 17$, which is rather easy, and already explained in the previous section.

More remarkably, this procedure is successful for $F_3 = 257$, too.

7.1. FERMAT PRIMES

As always, with primitive root 3, one obtains

$$\sigma(3,0,0) = 1 \; \sigma(3,1,0) = 1 \; \sigma(3,1,1) = 1$$
$$\sigma(3,2,0) = 1 \; \sigma(3,2,1) = -1 \; \sigma(3,2,2) = 1 \; \sigma(3,2,3) = -1$$

$$\sigma(3,3,0) = 1 \; \sigma(3,3,1) = 1 \; \sigma(3,3,2) = 1 \; \sigma(3,3,3) = 1$$
$$\sigma(3,3,4) = 1 \; \sigma(3,3,5) = 1 \; \sigma(3,3,6) = -1 \; \sigma(3,3,7) = 1$$

$$\sigma(3,4,0) = 1 \; \sigma(3,4,1) = 1 \; \sigma(3,4,2) = 1 \; \sigma(3,4,3) = 1$$
$$\sigma(3,4,4) = 1 \; \sigma(3,4,5) = 1 \; \sigma(3,4,6) = -1 \; \sigma(3,4,7) = -1$$
$$\sigma(3,4,8) = -1 \; \sigma(3,4,9) = -1 \; \sigma(3,4,10) = 1 \; \sigma(3,4,11) = -1$$
$$\sigma(3,4,12) = 1 \; \sigma(3,4,13) = -1 \; \sigma(3,4,14) = -1 \; \sigma(3,4,15) = 1$$

$$\sigma(3,5,0) = 1 \; \sigma(3,5,1) = 1 \; \sigma(3,5,2) = -1 \; \sigma(3,5,3) = 1$$
$$\sigma(3,5,4) = 1 \; \sigma(3,5,5) = 1 \; \sigma(3,5,6) = 1 \; \sigma(3,5,7) = -1$$
$$\sigma(3,5,8) = -1 \; \sigma(3,5,9) = -1 \; \sigma(3,5,10) = -1 \; \sigma(3,5,11) = -1$$
$$\sigma(3,5,12) = 1 \; \sigma(3,5,13) = -1 \; \sigma(3,5,14) = -1 \; \sigma(3,5,15) = 1$$
$$\sigma(3,5,16) = -1 \; \sigma(3,5,17) = -1, \; \sigma(3,5,18) = -1 \; \sigma(3,5,19) = -1$$
$$\sigma(3,5,20) = -1 \; \sigma(3,5,21) = -1 \; \sigma(3,5,22) = 1 \; \sigma(3,5,23) = 1$$
$$\sigma(3,5,24) = -1 \; \sigma(3,5,25) = -1 \; \sigma(3,5,26) = 1 \; \sigma(3,5,27) = 1$$
$$\sigma(3,5,28) = 1 \; \sigma(3,5,29) = -1 \; \sigma(3,5,30) = 1 \; \sigma(3,5,31) = -1$$

$$\sigma(3,6,0) = 1 \; \sigma(3,6,1) = 1 \; \sigma(3,6,2) = 1 \; \sigma(3,6,3) = 1$$
$$\sigma(3,6,4) = -1 \; \sigma(3,6,5) = 1 \; \sigma(3,6,6) = 1 \; \sigma(3,6,7) = -1$$
$$\sigma(3,6,8) = -1 \; \sigma(3,6,9) = -1 \; \sigma(3,6,10) = -1 \; \sigma(3,6,11) = 1$$
$$\sigma(3,6,12) = 1 \; \sigma(3,6,13) = 1 \; \sigma(3,6,14) = 1 \; \sigma(3,6,15) = -1$$
$$\sigma(3,6,16) = 1 \; \sigma(3,6,17) = 1 \; \sigma(3,6,18) = 1 \; \sigma(3,6,19) = 1$$
$$\sigma(3,6,20) = -1 \; \sigma(3,6,21) = -1 \; \sigma(3,6,22) = -1 \; \sigma(3,6,23) = 1$$
$$\sigma(3,6,24) = 1 \; \sigma(3,6,25) = -1 \; \sigma(3,6,26) = 1 \; \sigma(3,6,27) = -1$$
$$\sigma(3,6,28) = 1 \; \sigma(3,6,29) = 1 \; \sigma(3,6,30) = -1 \; \sigma(3,6,31) = -1$$
$$\sigma(3,6,32) = -1 \; \sigma(3,6,33) = -1 \; \sigma(3,6,34) = -1 \; \sigma(3,6,35) = 1$$

$\sigma(3,6,36) = 1$ $\sigma(3,6,37) = -1$ $\sigma(3,6,38) = -1$ $\sigma(3,6,39) = -1$
$\sigma(3,6,40) = -1$ $\sigma(3,6,41) = -1$ $\sigma(3,6,42) = -1$ $\sigma(3,6,43) = -1$
$\sigma(3,6,44) = 1$ $\sigma(3,6,45) = 1$ $\sigma(3,6,46) = -1$ $\sigma(3,6,47) = -1$
$\sigma(3,6,48) = 1$ $\sigma(3,6,49) = 1$ $\sigma(3,6,50) = 1$ $\sigma(3,6,51) = 1$
$\sigma(3,6,52) = -1$ $\sigma(3,6,53) = 1$ $\sigma(3,6,54) = -1$ $\sigma(3,6,55) = 1$
$\sigma(3,6,56) = 1$ $\sigma(3,6,57) = 1$ $\sigma(3,6,58) = -1$ $\sigma(3,6,59) = -1$
$\sigma(3,6,60) = 1$ $\sigma(3,6,61) = -1$ $\sigma(3,6,62) = -1$ $\sigma(3,6,63) = -1$

From here it is routine to build the formula with the box square roots for the coordinates of the 257-gon vertices,—at least in principle!
For more convincing news, I take up mathematica.

7.1.4 The 257-gon with mathematica

This is a mathematica program that computates the numbers $e[3,q,d]$ and $\sigma[3,q,d]$ exactly; next computates numerically the Gaussian sums $S[3,q,d]$ by means of above obtained formulas

((7.1.37)) $\Delta(n,q,l) = S(n,q,l)^2 - 4 \sum_{0 \le d < 2^q} e[n,q,d] S(n,q,l+d)$

((7.1.38))
$$S(n, q+1, l) = \frac{S(n,q,l) + \sigma(n,q,l)\sqrt{\Delta(n,q,l)}}{2}$$

((7.1.39))
$$S(n, q+1, l+2^q) = \frac{S(n,q,l) - \sigma(n,q,l)\sqrt{\Delta(n,q,l)}}{2}$$

and finally checks that the values obtained for $S[3,7,d]$ with $d = 0, 1, \ldots, 127$ agree with equation (7.1.40)

$$2 \cos \frac{2\pi \cdot 3^l}{F_n} = S(n, 2^n - 1, l) \text{ for all } 0 \le l < h$$

by directly calculating $2\cos[\frac{2\pi i \cdot 3^r}{257}]$ for $r = 0, 1, \ldots, 127 = h-1$ and comparing.

```
In[1]:= n = 3; Fermat = Function[n, 2^(2^n) + 1];
  Fermat[n]
Out[1]= 257
```

7.1. FERMAT PRIMES 213

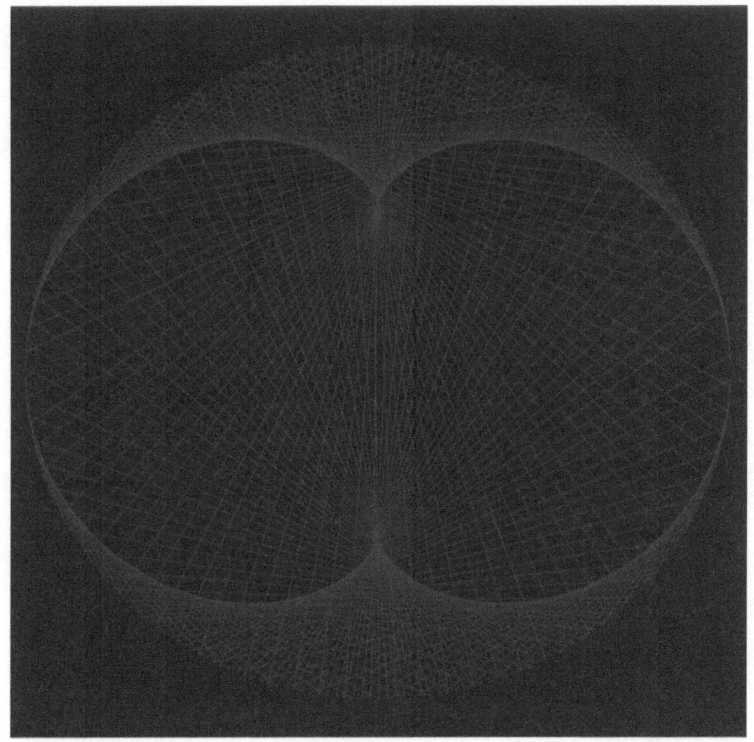

Figure 7.1.1: Vertices of a regular 257-gon connected as in $r \mapsto \exp[\frac{2\pi i \cdot 3^r}{257}]$

```
In[2]:= proot = PrimitiveRootList[Fermat[n]][[1]]
Out[2]= 3

In[3]:= Gauss = Function[{p, proot, d, s},
   e = (p - 1)/d;
   If[IntegerQ[e] && PrimeQ[p],
 ars = PowerMod[proot, Array[s+(# - 1)*d &, e ], p];
 eulersp = Function[t,
       exponent = 2*Pi *I*t/p;
       Exp[exponent]] ;
     Total[Map[eulersp, ars]],
     {"Nonprime or Noninteger", p, e} ]];

In[4]:= Gaussums = Function[{p, proot, d},
   Array[Gauss[ p, proot, d, #] &, d, 0]] ;

In[5]:= vertices = Gaussums[Fermat[n],proot,2^(2^n)];

In[6]:= ListLinePlot[
    (Tooltip[{Re[#1], Im[#1]}] &)/@ vertices,
 PlotStyle -> Directive[Hue[0.67, 0.6, 0.6],
 AbsoluteThickness[0.1]],
    AspectRatio -> 1, Axes -> False]
In[7]:= Clear[sigma]; sigma[q_, 1_]
    := sigma[q, 1] =
  Sign[Re[N[Gauss[Fermat[n], proot, 2^(q + 1), 1] -
        Gauss[Fermat[n], proot, 2^(q + 1), 1 + 2^q]]]]

In[8]:= baum = Array[Function[q,
        Array[{q, #} &, 2^q, 0]], 2^n, 0];

In[9]:= Clear[ex0]; ex0[q_, d_] := ex0[q, d] = With[{
    nq = 2^(2^n - q - 1),
    aq = PowerMod[3, 2^q, Fermat[n]]},
    count = 0; For[y = 0, y < nq, y++,
    try = 1 + PowerMod[aq, 2*y + 1, Fermat[n]] -
      PowerMod[3, d, Fermat[n]];
    If[Divisible[try, Fermat[n]], count++]]; count]

In[10]:= Clear[exsn]; exsn[q_, d_] := exsn[q, d] =
  Which[q == 0 , 2^(2^n - 2),
    q == 1 , 2^(2^n - 4),
```

7.1. FERMAT PRIMES 215

```
    q >= 2 && q <= 2^n - 2,
    With[{nq = 2^(2^n - q - 1)},
      Sum[exO[q, d + x*2^(q + 1)], {x, 0, nq - 1}]],
    True, Indeterminate]

In[11]:=Map[Function[pair,exsn[pair[[1]],pair[[2]]]],
 Most[baum], {2}]

Out[11]= {{64}, {16, 16}, {2, 5, 4, 5},
 {2, 0, 2, 0, 1, 2, 1, 0},
 {1,1,1,0,0,1,0,0,0,0,0,0,0 0,0,0},
 {0, 1, 0, 0, 0, 0, 0,  0, 0, 0, 0,
   0, 0, 0, 0, 0, 0, 0, 0, 0, 0, 0, 0, 1,
   0, 0, 0, 0, 0, 0, 0, 0},
  {0, 0, 0, 0, 0, 0, 0, 0, 0, 0, 0,
     0, 0, 0, 0, 0, 0, 0, 0, 0,
   0, 0, 0, 0, 0, 0, 0, 0, 0, 0,
     0, 0,0, 0, 0, 0, 0, 0, 0,
  0, 0, 0, 0, 0, 0, 0, 0, 0, 0,
    0, 0, 0, 0, 0, 0,1,0,0,
      0, 0, 0, 0, 0}}

In[12]:= exsn[6, 56]
Out[12]= 1

In[13]:= Array[Function[q,Sum[exsn[q, d],
    {d, 0, 2^q - 1, 1}]],
 2^n - 1, 0]

Out[13]= {64, 32, 16, 8, 4, 2, 1}

In[14]:= Array[Function[q,4 Sum[exsn[q,zweid],
    {zweid, 0, 2^q - 1, 2}]], 2^n - 1, 0]
Out[14]= {256, 64, 24, 24, 8, 0, 4}

In[15]:= Array[
 Function[q,
   2^(2^n - q - 1) +
    Sum[JacobiSymbol[1
        + PowerMod[3, (2 y + 1)*2^q, Fermat[n]],
      Fermat[n]],
    {y, 0, 2^(2^n - q - 1) - 1}]], 2^n - 1, 0]
```

Out[15]= {128, 64, 24, 24, 8, 0, 4}

```
In[16]:= Clear[nextS0];
   nextS0 = Function[{q, l, S, mysigma},
   P = S[q, l];
Q = N[Sum[exsn[q,d]*S[q,Mod[l+d,2^q]],{d,0,2^q-1}]];
   solutions=x /. NSolve[x^2-P*x+Q == 0,x,Reals];
   If[mysigma == -1, solutions[[1]], solutions[[2]] ]];

In[17]:= nextS = Function[{q, l, S}, Which[
   q < 2^n && l >= 0 && l < 2^q,
                nextS0[q, l, S, sigma[q, l]],
   q < 2^n && l >= 2^q && l < 2^(q + 1),
   nextS0[q, l - 2^q, S, -1*sigma[q, l - 2^q]],
   True, Indeterminate]];

In[18]:= ClearAll[S]; S[q_, l_] := S[q, l] =
   Which[q == 0 && l == 0, -1,
   q > 0 && q < 2^n, nextS[q - 1, l, S],
   True, Indeterminate]

In[19]:=Map[Function[pair,sigma[pair[[1]],pair[[2]]]],
Most[baum], {2}]

Out[19]= {{1}, {1, 1}, {1, -1, 1, -1},
 {1, 1, 1, 1, 1, 1, -1, 1},
 {1,1,1,1,1,1, -1, -1, -1, -1, 1,-1,1,-1,-1,1},
  {1, 1, -1, 1, 1, 1, 1, -1, -1, -1, -1, -1,
   1, -1, -1, 1, -1, -1, -1, -1, -1, -1,1, 1,
   -1, -1, 1, 1, 1, -1, 1, -1},
  {1, 1, 1, 1, -1, 1,1, -1, -1, -1,
    -1, 1, 1, 1, 1, -1, 1, 1, 1, 1,
   -1, -1, -1, 1, 1, -1,1, -1, 1, 1,
    -1, -1, -1, -1, -1, 1, 1, -1, -1,
    -1, -1, -1, -1, -1,1, 1, -1, -1,
    1, 1, 1, 1, -1,
    1, -1, 1, 1, 1,-1, -1, 1, -1, -1, -1}}

In[20]:=Map[Function[pair,S[pair[[1]],pair[[2]]]],
   Take[baum, 4], {2}];
```

7.1. FERMAT PRIMES 217

```
In[21]:=Map[Function[pair, S[pair[[1]],pair[[2]]]],
   Take[baum, 6], {2}];

In[22]:= NSbaum=Map[Function[
 pair, S[pair[[1]], pair[[2]]]], baum, {2}];

In[23]:= zweicos = NSbaum[[8]];

In[24]:= Ndef = Function[
 cs2,vert=Fermat[n]*ArcCos[cs2/2]/(2*Pi);
   vert - Round[vert]];

In[25]:= Max[Map[Ndef, zweicos]] ;;
            Min[Map[Ndef, zweicos]]
Out[25]= 1.9061*10^-10 ;; -3.91367*10^-10

In[26]:= Nvert =  Function[
 cs2, vert = Fermat[n]*ArcCos[cs2/2]/(2 Pi);
                        Round[vert]];

In[27]:= waslos = Map[Nvert, zweicos]
Out[27]= {1, 3,9,27,81,14,42,126,121,106,61,74,35,
 105,58,83,8,24,72,41,123,112,79,20,60,77,26,78,23,
 69,50,107,64,65,62,71,44,125,118,97,34,102,49,110,
 73,38,114,85,2,6,18,54,95,28,84,5,15,45,122,109,
  70,47,116,91,16,48,113,82,11,33,99,40,120,103,52,
  101,46,119,100,43,128,127,124,115,88,7,21,63,68,
  53,98,37,111,76,29,87,4,12,36,108,67,56,89,10,30,
  90,13,39,117,94,25,75,32,96,31,93,22,66,59,80,17,
  51,104,55,92,19,57,86}

In[28]:= twoxvertices=Gaussums[
   Fermat[n],proot, 2^(2^n-1)];

In[29]:=  easy = Map[Fermat[n]*ExpToTrig[
 ArcCos[#/2]]/(2 Pi) &, twoxvertices]
Out[29]= {1, 3, 9, 27, 81, 14, 42, 126,121,106,
  61, 74, 35,105,58, 83, 8, 24, 72, 41, 123, 112,
   79, 20, 60, 77, 26,78,23,69,50,107, 64, 65, 62,
   71, 44, 125, 118, 97, 34,102, 49,110,73,38,
   114,85, 2, 6, 18, 54, 95, 28, 84, 5,15,45,
   122, 109, 70, 47, 116, 91,16, 48, 113, 82, 11,
```

33,99, 40, 120,103,52,101,46,119,100,43,
128, 127,124,115, 88, 7, 21, 63, 68, 53,98,
37,111, 76, 29, 87,4,12,36,108,67,56,89,
10,30,90,13,39,117,94,25,75,32,96,31,93,
22, 66,59,80,17,51,104, 55,92,19,57,86}

```
In[30]:= waslos == easy
Out[30]= True

In[31]:= GN = Array[Map[Re[N[#, 20]] &,
    Gaussums[Fermat[n], proot, 2^#]] &, 2^n,0];

In[32]:= Max[GN-NSbaum] ;; Min[GN-NSbaum]
Out[32]= 9.57412*10^-12 ;; -9.77995*10^-12

In[33]:= Map[Ndef, NSbaum[[8]]];
```

What about $F_4 = 65\,537$?

7.1.5 Computations for the $65\,537$-gon

The first programming attempt with DrRacket did not succeed because of memory problems. One needs indeed produce the coefficients $e[4, q, d]$ with *one counting procedure* derived from equation (7.1.30), and has to avoid keeping any intermediate results. In this manner the amount of computer time needed can be managed. Above I have given only the program written according to this requirement. Now I finally am going to write down the nonzero coefficients:

$$e[4, 0, 0] = 16384,$$
$$e[4, 1, 0] = 4096, e[4, 1, 1] = 4096,$$
$$e[4, 2, 0] = 992, e[4, 2, 1] = 1040, e[4, 2, 2] = 1024, e[4, 2, 3] = 1040,$$
$$e[4, 3, 0] = 284, e[4, 3, 1] = 237, e[4, 3, 2] = 272, e[4, 3, 3] = 237,$$
$$e[4, 3, 4] = 256, e[4, 3, 5] = 269, e[4, 3, 6] = 256, e[4, 3, 7] = 237,$$

7.1. FERMAT PRIMES

$e[4,4,0] = 80, e[4,4,1] = 62, e[4,4,2] = 60, e[4,4,3] = 64,$
$e[4,4,4] = 57, e[4,4,5] = 60, e[4,4,6] = 61, e[4,4,7] = 60,$
$e[4,4,8] = 68, e[4,4,9] = 64, e[4,4,10] = 64, e[4,4,11] = 58,$
$e[4,4,12] = 65, e[4,4,13] = 70, e[4,4,14] = 61, e[4,4,15] = 70,$
$e[4,5,0] = 4, e[4,5,1] = 12, e[4,5,2] = 20, e[4,5,3] = 13,$
$e[4,5,4] = 20, e[4,5,5] = 18, e[4,5,6] = 16, e[4,5,7] = 19,$
$e[4,5,8] = 19, e[4,5,9] = 22, e[4,5,10] = 12, e[4,5,11] = 22,$
$e[4,5,12] = 13, e[4,5,13] = 13, e[4,5,14] = 11, e[4,5,15] = 22,$
$e[4,5,16] = 20, e[4,5,17] = 15, e[4,5,18] = 25, e[4,5,19] = 12,$
$e[4,5,20] = 16, e[4,5,21] = 12, e[4,5,22] = 16, e[4,5,23] = 17,$
$e[4,5,24] = 29, e[4,5,25] = 16, e[4,5,26] = 7, e[4,5,27] = 17,$
$e[4,5,28] = 13, e[4,5,29] = 17, e[4,5,30] = 13, e[4,5,31] = 11,$

$e[4,6,1] = 3, e[4,6,2] = 2, e[4,6,3] = 5, e[4,6,4] = 5, e[4,6,5] = 5,$
$e[4,6,6] = 2, e[4,6,7] = 5, e[4,6,8] = 2, e[4,6,9] = 6, e[4,6,10] = 5,$
$e[4,6,11] = 6, e[4,6,12] = 1, e[4,6,13] = 6, e[4,6,14] = 3,$
$e[4,6,15] = 5, e[4,6,16] = 6, e[4,6,17] = 8, e[4,6,18] = 3,$
$e[4,6,19] = 5, e[4,6,20] = 2, e[4,6,21] = 5, e[4,6,22] = 5,$
$e[4,6,23] = 1, e[4,6,24] = 3, e[4,6,25] = 10, e[4,6,26] = 3,$
$e[4,6,27] = 3, e[4,6,28] = 4, e[4,6,29] = 5, e[4,6,30] = 1,$
$e[4,6,31] = 4, e[4,6,32] = 1, e[4,6,33] = 6, e[4,6,34] = 3,$
$e[4,6,35] = 2, e[4,6,36] = 1, e[4,6,37] = 1, e[4,6,38] = 7,$
$e[4,6,39] = 3, e[4,6,40] = 3, e[4,6,41] = 3, e[4,6,42] = 4,$
$e[4,6,43] = 6, e[4,6,44] = 3, e[4,6,45] = 2, e[4,6,46] = 3,$
$e[4,6,47] = 4, e[4,6,48] = 7, e[4,6,49] = 7, e[4,6,50] = 4,$
$e[4,6,51] = 4, e[4,6,52] = 5, e[4,6,53] = 6, e[4,6,54] = 8,$
$e[4,6,55] = 2, e[4,6,56] = 2, e[4,6,57] = 4, e[4,6,58] = 3,$
$e[4,6,59] = 4, e[4,6,60] = 3, e[4,6,61] = 5, e[4,6,62] = 8,$
$e[4,6,63] = 3,$

$e[4,7,0] = 2, e[4,7,1] = 1, e[4,7,3] = 1, e[4,7,4] = 1, e[4,7,6] = 1,$
$e[4,7,7] = 1, e[4,7,8] = 1, e[4,7,10] = 3, e[4,7,14] = 1,$
$e[4,7,15] = 1, e[4,7,16] = 1, e[4,7,20] = 1, e[4,7,22] = 1,$
$e[4,7,25] = 1, e[4,7,26] = 1, e[4,7,29] = 3, e[4,7,30] = 1,$
$e[4,7,33] = 2, e[4,7,34] = 2, e[4,7,35] = 1, e[4,7,36] = 2,$
$e[4,7,37] = 1, e[4,7,38] = 2, e[4,7,40] = 2, e[4,7,41] = 2,$
$e[4,7,42] = 2, e[4,7,45] = 1, e[4,7,47] = 1, e[4,7,48] = 1,$
$e[4,7,51] = 1, e[4,7,56] = 1, e[4,7,57] = 5, e[4,7,58] = 1,$
$e[4,7,59] = 2, e[4,7,60] = 1, e[4,7,61] = 1, e[4,7,63] = 2,$
$e[4,7,65] = 4, e[4,7,67] = 2, e[4,7,70] = 2, e[4,7,73] = 2,$
$e[4,7,74] = 2, e[4,7,76] = 2, e[4,7,79] = 2, e[4,7,80] = 1,$
$e[4,7,82] = 1, e[4,7,83] = 2, e[4,7,86] = 1, e[4,7,87] = 2,$
$e[4,7,88] = 2, e[4,7,89] = 1, e[4,7,90] = 1, e[4,7,94] = 2,$
$e[4,7,95] = 1, e[4,7,97] = 2, e[4,7,98] = 1, e[4,7,99] = 1,$
$e[4,7,100] = 3, e[4,7,101] = 2, e[4,7,103] = 2, e[4,7,104] = 2,$
$e[4,7,105] = 3, e[4,7,106] = 4, e[4,7,108] = 2, e[4,7,109] = 2,$
$e[4,7,110] = 1, e[4,7,111] = 1, e[4,7,114] = 4, e[4,7,115] = 1,$
$e[4,7,116] = 1, e[4,7,117] = 2, e[4,7,118] = 1, e[4,7,119] = 1,$
$e[4,7,121] = 2, e[4,7,123] = 1, e[4,7,124] = 1,$
$e[4,7,125] = 1, e[4,7,127] = 2,$

7.1. FERMAT PRIMES 221

$e[4, 8, 3] = 2, e[4, 8, 4] = 1, e[4, 8, 7] = 1, e[4, 8, 14] = 1,$
$e[4, 8, 19] = 1, e[4, 8, 28] = 2, e[4, 8, 29] = 1, e[4, 8, 30] = 1,$
$e[4, 8, 33] = 1, e[4, 8, 37] = 1, e[4, 8, 39] = 1, e[4, 8, 41] = 1,$
$e[4, 8, 43] = 1, e[4, 8, 44] = 1, e[4, 8, 50] = 1, e[4, 8, 51] = 1,$
$e[4, 8, 53] = 1, e[4, 8, 56] = 1, e[4, 8, 59] = 1, e[4, 8, 61] = 1,$
$e[4, 8, 65] = 1, e[4, 8, 68] = 1, e[4, 8, 70] = 1, e[4, 8, 78] = 1,$
$e[4, 8, 79] = 1, e[4, 8, 81] = 1, e[4, 8, 82] = 3, e[4, 8, 84] = 1,$
$e[4, 8, 88] = 1, e[4, 8, 89] = 1, e[4, 8, 106] = 1, e[4, 8, 109] = 1,$
$e[4, 8, 112] = 1, e[4, 8, 117] = 1, e[4, 8, 124] = 1, e[4, 8, 128] = 1,$
$e[4, 8, 135] = 1, e[4, 8, 142] = 1, e[4, 8, 146] = 1, e[4, 8, 156] = 2,$
$e[4, 8, 173] = 1, e[4, 8, 175] = 1, e[4, 8, 185] = 2, e[4, 8, 186] = 1,$
$e[4, 8, 187] = 1, e[4, 8, 188] = 1, e[4, 8, 195] = 1, e[4, 8, 200] = 1,$
$e[4, 8, 212] = 1, e[4, 8, 219] = 1, e[4, 8, 231] = 1, e[4, 8, 233] = 1,$
$e[4, 8, 243] = 1, e[4, 8, 245] = 2,$
$e[4, 8, 250] = 1,$
$e[4, 8, 252] = 1, e[4, 8, 253] = 1,$

$e[4, 9, 40] = 1, e[4, 9, 48] = 1, e[4, 9, 67] = 2, e[4, 9, 80] = 1,$
$e[4, 9, 85] = 2, e[4, 9, 87] = 2, e[4, 9, 91] = 1, e[4, 9, 105] = 1,$
$e[4, 9, 108] = 1, e[4, 9, 113] = 1, e[4, 9, 134] = 2, e[4, 9, 174] = 2,$
$e[4, 9, 210] = 1, e[4, 9, 225] = 1, e[4, 9, 232] = 1, e[4, 9, 268] = 1,$
$e[4, 9, 274] = 1, e[4, 9, 277] = 1, e[4, 9, 280] = 1, e[4, 9, 302] = 1,$
$e[4, 9, 348] = 1, e[4, 9, 378] = 1, e[4, 9, 389] = 1, e[4, 9, 430] = 1,$
$e[4, 9, 450] = 2, e[4, 9, 464] = 1,$

$e[4, 10, 0] = 2, e[4, 10, 23] = 1, e[4, 10, 154] = 2, e[4, 10, 184] = 1,$
$e[4, 10, 308] = 1, e[4, 10, 359] = 1, e[4, 10, 530] = 1,$
$e[4, 10, 666] = 1, e[4, 10, 718] = 1, e[4, 10, 733] = 1,$
$e[4, 10, 777] = 2, e[4, 10, 840] = 1, e[4, 10, 945] = 1,$

$$e[4,11,0] = 1, e[4,11,1] = 1, e[4,11,2] = 1,$$
$$e[4,11,777] = 1, e[4,11,800] = 1, e[4,11,1099] = 1,$$
$$e[4,11,1178] = 1, e[4,11,1263] = 1,$$
$$e[4,12,1] = 1, e[4,12,1265] = 1, e[4,12,1899] = 1,$$
$$e[4,12,4003] = 1, e[4,13,3164] = 1, e[4,13,8100] = 1,$$
$$e[4,14,3072] = 1,$$

I include in these recapitulation only the first few sign coefficients σ since I do not think they are really instructive.

$$\sigma(4,0,0) = 1 \quad \sigma(4,1,0) = -1 \quad \sigma(4,1,1) = -1$$
$$\sigma(4,2,0) = -1 \quad \sigma(4,2,1) = 1 \quad \sigma(4,2,2) = 1 \quad \sigma(4,2,3) = -1$$

Thus I may deduct that the construction of the 65 537-gon by straightedge and compass is in principle possible.

But one is still missing a serious check of all the above calculations! To this end, I can document more success using mathematica. Same as already for 257-gon, a complete calculation of the symbolic solutions for $2\cos\frac{2\pi l}{65\,537}$ is out of question.

Problem 99. *Estimate roughly how much memory and how much paper would be required.*

But I could at least achieve again a *numeric calculation*. With the standard 10 digit accuracy, the program dies at $q = 13$ because of the accumulation of rounding errors. I did use 35 digits, and I had to avoid the difference occuring in one of the quadratic formula (7.1.38) respectively (7.1.39) and use

$$S(n, q+1, l+2^q) = \frac{S(n, q+1, l) \cdot S(n, q+1, l+2^q)}{S(n, q+1, l)}$$
$$= \frac{\sum_{0 \le d < 2^q} e[n, q, d] S(n, q, l+d)}{S(n, q+1, l)}$$

With those tricks, the calculation went through to level $q = 2^n - 1 = 15$. Finally I could check the result by recalculation of the counting number l from the numerical value for $2\cos\frac{2\pi l}{65\,537}$:

```
N129 =  Map[Function[pair, NS[pair[[1]], pair[[2]]]],
                 teilbaum[16][[16]]];
Ndef = Function[cs2, vert
                 = 65537 *ArcCos[cs2/2]/(2 Pi);
   def = vert - Round[vert]; N[def]];
```

7.1. FERMAT PRIMES

```
Max[Map[Ndef, N129]]
0.000719844

Min[Map[Ndef, N129]]
-0.000725486
```

Hence we have obtained still 4 *accurate digits* for this monster calculation. These computations took about 6 hours. Now I am convinced that my construction of the regular 65 537-gon is correct.

7.1.6 Identities with Jacobi Symbols

Problem 100. *Let $0 \le q \le 2^n - 1$. Check that the following formula*

$$\left(\frac{1+9^{2y+1}}{F_n}\right) = \left(\frac{1+9^{2^{(2^n-1)}-(2y+1)}}{F_n}\right)$$

among Jacobi symbols holds for all $n \ge 1$ and all integer y such that $0 \le 2y+1 \le 2^{2^n-1}$.

Solution.

$$\left(\frac{1+9^{2^{(2^n-1)}-(2y+1)}}{F_n}\right) = \left(\frac{9^{2y+1}}{F_n}\right) \cdot \left(\frac{1+9^{2^{(2^n-1)}-(2y+1)}}{F_n}\right)$$

$$= \left(\frac{9^{2y+1}+9^{2^{(2^n-1)}}}{F_n}\right)$$

$$= \left(\frac{9^{2y+1}+3^{F_n-1}}{F_n}\right) = \left(\frac{1+9^{2y+1}}{F_n}\right)$$

$\square$

Lemma 82. *We show at first that $1 + 3^{(1+2y) \cdot 2^q}$ is not divisible by F_n for any $0 \le y < 2^{2^n-q-1}$ and $0 \le q \le 2^n - 2$.*

Proof. Assume the claim is wrong. Since 3 is a primitive root modulo F_n, one would get the congruences

$$1 + 3^{(1+2y) \cdot 2^q} \equiv 0 \pmod{F_n}$$
$$3^{(1+2y) \cdot 2^q} \equiv 3^{2^{2^n-1}} \pmod{F_n}$$
$$(1+2y) \cdot 2^q \equiv 2^{2^n-1} \pmod{2^{2^n}}$$
$$1 + 2y \equiv 2^{2^n-1-q} \pmod{2^{2^n-q}}$$

Since $1 \leq 1+2y \leq 2^{2^n-q}$ is assumed, the last line implies $1+2y = 2^{2^n-1-q}$. This is only possible for $q = 2^n - 1$ and $y = 0$, a case excluded by the assumptions. $\square$

Assume $n \geq 1$ and $0 \leq q \leq 2^n - 2$. Because of Lemma 82 these characteristic functions may be written in terms of a Jacobi symbol.

$$\sum_{0 \leq d < F_n - 1 \text{ and } d \text{ even}} \chi(t[0, y, q, d] \equiv 0 \pmod{F_n})$$

$$\sum_{0 \leq d < F_n - 1 \text{ and } d \text{ even}} \left[\chi(1 + 3^{(1+2y) \cdot 2^q} \equiv 3^d \pmod{F_n}) \right]$$

$$= \frac{1}{2} \left[1 + \left(\frac{1 + 3^{(1+2y) \cdot 2^q}}{F_n} \right) \right]$$

The formula (7.1.34) now implies

$$ex0[n, q, d] =$$
$$\sum_{0 \leq y < 2^{2^n-q-1}} \chi(t[0, y, q, d] \equiv 0 \pmod{F_n})$$
$$\sum_{0 \leq d < F_n - 1 \text{ and } d \text{ even}} ex0[n, q, d] =$$
$$\sum_{0 \leq y < 2^{2^n-q-1}} \frac{1}{2} \left[1 + \left(\frac{1 + 3^{(1+2y) \cdot 2^q}}{F_n} \right) \right]$$

From Lemma 79, and using that the function $d \mapsto ex0[n, q, d]$ has the period 2^{2^n} and the function $d \mapsto e[n, q, d]$ has the period 2^q,

7.1. FERMAT PRIMES

we get

$$e[n,q,d] = \sum_{0 \le x < 2^{2^n-q-1}} ex0[n,q,d+x*2^{q+1}]$$

$$\sum_{0 \le d < F_n-1 \text{ and } d \text{ even}} e[n,q,d] =$$

$$\left[\sum_{0 \le d < F_n-1 \text{ and } d \text{ even}} ex0[n,q,d+x*2^{q+1}]\right]$$

$$\sum_{0 \le d < F_n-1 \text{ and } d \text{ even}} e[n,q,d] =$$

$$\sum_{0 \le x < 2^{2^n-q-1}} \left[\sum_{0 \le d < F_n-1 \text{ and } d \text{ even}} ex0[n,q,d]\right]$$

$$= \sum_{0 \le x < 2^{2^n-q-1}} \left[\sum_{0 \le y < 2^{2^n-q-1}} \frac{1}{2}\left[1 + \left(\frac{1 + 3^{(1+2y) \cdot 2^q}}{F_n}\right)\right]\right]$$

$$2^{2^n-q} \sum_{0 \le d < 2^q \text{ and } d \text{ even}} e[n,q,d]$$

$$= 2^{2^n-q-1} \sum_{0 \le y < 2^{2^n-q-1}} \frac{1}{2}\left[1 + \left(\frac{1 + 3^{(1+2y) \cdot 2^q}}{F_n}\right)\right]$$

$$4 \sum_{0 \le d < 2^q \text{ and } d \text{ even}} e[n,q,d]$$

$$= 2^{2^n-q-1} + \sum_{0 \le y < 2^{2^n-q-1}} \left(\frac{1 + 3^{(1+2y) \cdot 2^q}}{F_n}\right)$$

$$4 \sum_{0 \le d < 2^q \text{ and } d \text{ odd}} e[n,q,d]$$

$$= 2^{2^n-q-1} - \sum_{0 \le y < 2^{2^n-q-1}} \left(\frac{1 + 3^{(1+2y) \cdot 2^q}}{F_n}\right)$$

Remark. I have checked the last two formulas with mathematica to be correct for $n = 3$ and $1 \le q < 7$. I do not understand the computational problem for $q = 0$ since the formula holds in this case, too.

The following checks turn out in agreement with Corollary 31 below. Especially we put $q = 2^n - 2 = 2$. For $n = 2$ one gets

$$4\chi \, (d^* \text{ is even}) = 2 + \sum_{0 \le y < 2} \left(\frac{1 + 3^{(1+2y) \cdot 2^{2^n - 2}}}{F_n} \right)$$

$$= 2 + \left(\frac{1 + 3^{(2^{2^n - 2})}}{F_n} \right) + \left(\frac{1 + 3^{3 \cdot (2^{2^n - 2})}}{F_n} \right)$$

$$= 2 + \left(\frac{14}{17} \right) + \left(\frac{5}{17} \right) = 0$$

Hence d^* is odd. Especially, put $n = 3$ and $q = 2^n - 2$.

$$4\chi \, (d^* \text{ is even}) = 2 + \left(\frac{1 + 3^{(2^{2^n - 2})}}{F_n} \right) + \left(\frac{1 + 3^{3 \cdot (2^{2^n - 2})}}{F_n} \right)$$

$$= 2 + \left(\frac{242}{257} \right) + \left(\frac{17}{257} \right) = 4$$

Hence d^* is even. Especially, put $n = 4$ and $q = 2^n - 2$.

$$4\chi \, (d^* \text{ is even}) =$$

$$2 + \left(\frac{1 + 3^{(2^{2^n - 2})}}{F_n} \right) + \left(\frac{1 + 3^{3 \cdot (2^{2^n - 2})}}{F_n} \right)$$

$$= 2 + \left(\frac{65282}{65537} \right) + \left(\frac{257}{65537} \right) = 4$$

Hence d^* is even.

Now I turn to the case $q = 1$.

$$4e[n, 1, 0] = 4 \sum_{0 \le d < 2^q \text{ and } d \text{ even}} e[n, q, d]$$

$$= \sum_{0 \le y < 2^{2^n - 2}} \left[1 + \left(\frac{1 + 9^{1 + 2y}}{F_n} \right) \right]$$

$$e[n, 1, 0] = 2^{2^n - 4} + \frac{1}{4} \sum_{0 \le y < 2^{2^n - 2}} \left(\frac{1 + 9^{1 + 2y}}{F_n} \right)$$

$$e[n, 1, 1] = 2^{2^n - 4} - \frac{1}{4} \sum_{0 \le y < 2^{2^n - 2}} \left(\frac{1 + 9^{1 + 2y}}{F_n} \right)$$

Actually, the latter sum of Jacobi symbols is zero for $n = 2, 3, 4$. I could not prove this in general.

7.1. FERMAT PRIMES 227

Problem 101. *Deduct from the above result with $q = 0$ the following sum of Jacobi symbols*

(7.1.41) $$\sum_{0 \leq y < 2^{2^n-1}} \left(\frac{1 + 3^{(1+2y)}}{F_n} \right) = 0$$

Problem 102. *Prove the following formula*

$$\sum_{0 \leq y < 2^{2^n-2}} \left(\frac{1 + 9^{1+2y}}{F_n} \right) = 0$$

among Jacobi symbols holds for all $n \geq 1$. I could get the formula only for $1 \leq n \leq 4$.

Problem 103. *Prove that $e[n, 1, 0] = e[n, 1, 1] = 2^{2^n - 4}$ holds,— not only for $n = 2, 3, 4$ as we shall check numerically,—but for all n for which F_n turns out to be a Fermat prime.*

Chapter 8

High Fermat numbers

8.0.1 More about Fermat Numbers

Problem 104. *For $n \geq 3$, each primitive root is a quadratic non-residue. But the converse is very rare. Prove that for $n \geq 3$, each quadratic non-residue is a primitive root if and only if the number is either $n = 4$ or $n = p$ or $n = 2p$ where p is a Fermat prime. For example, $n = 10$ has the primitive root 3. its powers make the unit group: $\{3, 9, 7, 1\} = \mathcal{U}(\mathbf{Z}_{10})$. The primitive roots,—and the non-residues as well,—are 3 and 7.*

Distinguish the cases

(o) $n = 4$.

(a) *A primitive root exists and $n = p$ or $n = 2p$ where p is an odd prime.*

(b) *A primitive root exists and $n = p^s$ or $n = 2p^s$ where p is an odd prime and $s \geq 2$.*

(c) *No primitive root exists.*

Remark. For $n = 2$ there exists the unique primitive root 1. But no quadratic non-residue exists. It is vacuously true that each quadratic non-residue is a primitive root.

Solution. **(o4)** For $n = 4$ there exists the unique primitive root 3. Too, 3 is the unique quadratic non-residue. Each quadratic non-residue is a primitive root.

229

(a) A primitive root exists and $n = p$ or $n = 2p$ where p is an odd prime. Let a be a primitive root. By lemma 46 the number of primitive roots is $\phi(\phi(n))$ if there exists a primitive root. The quadratic nonresidues are the odd powers $a, a^3, \ldots, a^{\phi(n)-1}$. Hence there exist $\frac{\phi(n)}{2}$ quadratic non-residues. Hence

$$\frac{\phi(n)}{2} = \phi(\phi(n)) \text{ and } \frac{p-1}{2} = \phi(p-1)$$

Hence $p-1$ is a power of 2. By lemma 68 we know: if p is any odd prime, and $p-1$ is a power of two, then p is a Fermat prime. Indeed, each quadratic non-residue is a primitive root if and only if n is a Fermat prime or its double.

(b) A primitive root exists and $n = p^s$ or $n = 2p^s$ where p is an odd prime. Take the case $s \geq 2$, which indeed leads to a contradiction. The number of primitive roots is

$$\phi(\phi(n)) = \phi(p^{s-1}(p-1)) = \phi(p^{s-1})\phi(p-1)$$
$$= p^{s-2}(p-1)\phi(p-1)$$

Let $p - 1 = 2^t \cdot u$ where $t \geq 1$ and u is odd. We conclude

$$\frac{\phi(n)}{2} = \phi(\phi(n))$$
$$p^{s-1}\frac{p-1}{2} = p^{s-2}(p-1)\phi(p-1)$$
$$p = 2\phi(p-1)$$
$$1 + 2^t u = 2^t \phi(u)$$

which is a contradiction. It cannot happen that each quadratic non-residue is a primitive root.

(c) No primitive root exists. Hence $n \neq 2$. By proposition 43, two.is the only number, for which no quadratic non-residues exist. It cannot happen that each quadratic non-residue is a primitive root. $\square$

Problem 105. *Prove by induction that $F_0 \cdot F_1 \cdots F_{n-1} = F_n - 2$ for all natural numbers $n \geq 1$.*

Answer. **Basic step:** For $n = 1$, both sides of the formula are equal to 3, since $3 = F_0 = F_1 - 2 = 5 - 2$.

Induction step "$n \to n+1$": The formula is assumed to hold for n as written. It is shown for n replaced by $n+1$, as follows:

$$F_0 \cdot F_1 \cdots F_n = [F_0 \cdot F_1 \cdots F_{n-1}] \cdot F_n \quad \text{recursive define product}$$
$$= (F_n - 2)F_n \quad \text{induction assumption}$$
$$= \left(2^{2^n} - 1\right)\left(2^{2^n} + 1\right) = \left(2^{2^n}\right)^2 - 1^2 = 2^{2^{n+1}} - 1$$
$$= F_{n+1}$$

We have checked the asserted formula step for $n+1$. From the basic step and the induction step together, we conclude by the principle of induction that the formula holds for all $n \geq 1$.

Problem 106. *Use the last problem to conclude Goldbach's Theorem, which states that any two different Fermat numbers are relatively prime. Conclude there exist infinitely many primes.*

Answer. Assume that p is a common prime factor of the Fermat numbers F_k and F_n with $0 \leq k < n$. We could conclude that p is a divisor of $F_0 \cdot F_1 \cdots F_{n-1} = F_n - 2$, and hence of both $F_n - 2$ and of F_n. This is only possible for $p = 2$. But all Fermat numbers are odd and cannot have divisor 2. This argument excludes that any two Fermat numbers have a common prime factor.

For any natural number n, let p_n be the smallest (or any) prime factor of F_n. The primes p_n are all different since the Fermat numbers are relatively prime. Hence there exist infinitely many primes.

Corollary 32. *In any arithmetic sequence $1 + i\, 2^k$ with $i \geq 1$ there exist infinitely many primes.*

Reason. Any prime factor of F_n has the form $p_n = 1 + i\, 2^{n+2}$. Hence all p_n with $n \geq k-2$ are contained in the arithmetic sequence $1 + i\, 2^k$ with $i = 1, 2, 3, \ldots$. □

Proposition 63 (Euler 1770). *Any prime factor of the Fermat number F_n has the form $p_n = 1 + i\, 2^{n+1}$.*

Reason. Assume that p is a prime factor of the Fermat number F_n. Let $h > 1$ be the smallest integer such that $2^h \equiv 1 \mod p$. This power h is also called the *order* of 2 modulo p. The definition of F_n implies

$$2^{2^n} \equiv -1 \mod F_n \quad \text{and} \quad 2^{2^{n+1}} \equiv 1 \mod F_n$$
$$2^{2^n} \equiv -1 \mod p \quad \text{and} \quad 2^{2^{n+1}} \equiv 1 \mod p$$

Since h is a divisor of 2^{n+1} but not 2^n, we get $h = 2^{n+1}$.

Fermat's Little Theorem implies $2^{p-1} \equiv 1 \mod p$, and hence h is a divisor of $p - 1$. Together we conclude that 2^{n+1} is a divisor of $p - 1$, as to be shown. □

It took more than hundred years, until Édouard Lucas improved Euler's result.

Proposition 64 (Lucas 1878). *Any prime factor of the Fermat number F_n with $n \geq 2$ has the form $p_n = 1 + j\, 2^{n+2}$.*

Reason. We continue the reasoning from Euler's Proposition 63. Assume that p is a prime factor of the Fermat number F_n and $n \geq 2$. As a consequence of Euler's Proposition, we see that 8 is a divisor of $p - 1$. Hence one can calculate the Legendre symbol

$$\left(\frac{2}{p}\right) = (-1)^{\frac{(p-1)(p+1)}{8}} = 1$$

and one gets from Euler's criterium

$$2^{\frac{p-1}{2}} \equiv \left(\frac{2}{p}\right) = 1 \mod p$$

Hence the *order* h of 2 modulo p is actually a divisor of $(p-1)/2$. We know that $h = 2^{n+1}$ and conclude that 2^{n+2} is a divisor of $p - 1$, as to be shown. □

Problem 107. *Explain why any number $N = 1 + 2^k$ is either a Fermat number or is divisible by a Fermat number.*

Proof. Let $k = 2^n \cdot (2b + 1)$. One may factor

$$N = 1 + 2^{2^n \cdot (2b+1)} = (1 + 2^{2^n})(1 - 2^{2^n} + 2^{2^n \cdot 2} - \cdots + 2^{2^n \cdot 2b})$$

Hence N is divisible by the Fermat number F_n. □

Corollary 33. *Any proper factor of the Fermat number F_n with $n \geq 5$ has the form $p_n = 1 + j\, 2^{n+2}$ with $j \geq 3$ and j not a power of 2.*

Proof. Any prime factor of F_n cannot be equal to $K = 1 + j\, 2^{n+2}$ with $j = 1$ or j a power of two since the number K would be a Fermat prime, but the Fermat numbers are relatively prime to each other.

Any proper factor of F_n cannot be equal to $1 + j\, 2^{n+2}$ with $j = 1$ or j a power of two since such a number would have a Fermat prime as a prime factor. This is impossible since the Fermat numbers are relatively prime to each other. □

Remark. It turns out occur many times that the Fermat number F_n has a prime factor $1 + j\, 2^{n+2}$ with j odd. This is known to happen for

$$n = 5, 6, 7, 9, 10, 11, 12, 15, 17, 18, 19, 21, 23, 25, 27, 29, 30, 31, 32$$

and many more cases with $n \geq 33$. In this sense, Lucas' result turns out to be optimal. On the other hand, F_8 has no prime factor with j odd.

Theorem 22 (Pépin's test (1877)). *For $n \geq 1$, the Fermat number F_n is prime if and only if*

$$3^{(F_n-1)/2} \equiv -1 \mod F_n$$

In 1905 and 1909, J.C. Morehead and A.E. Western used Pépins test to prove that F_7 and F_8 are composite. As of 2001, no factor is known for the Fermat numbers F_n with $n = 14, 20, 22, 24$. These numbers were proved composite only with Pépin's test. Here is a proof and some results related to this test.

Proposition 65 (Sufficiency of a Pépin-like test). *Assume there exists a natural number a such that*

$$a^{(F_n-1)/2} \equiv -1 \mod F_n$$

Then the Fermat number F_n is prime.

Proof. The assumption implies

(8.0.1) $\quad a^{(F_n-1)/2} \equiv -1 \mod F_n \quad \text{and} \quad a^{F_n-1} \equiv 1 \mod F_n$

Let p be any prime factor of the Fermat number F_n. Let $h > 1$ be the smallest integer such that $a^h \equiv 1 \mod p$. Since the congruences (8.0.1) hold modulo p, too, we see that h divides $F_n - 1$ but not $(F_n - 1)/2$, and hence $h = F_n - 1$. The Little Fermat Theorem implies

$$a^{p-1} \equiv 1 \mod p$$

and hence h is a divisor of $p - 1$. We conclude

$$1 + h = F_n \leq p \leq F_n$$

Hence $F_n = p$ is a prime, as to be shown. $\square$

Lemma 83. *No matter whether F_n is prime or not, the Jacobi symbols are*

$$\left(\frac{F_n}{a}\right) = \left(\frac{a}{F_n}\right) = -1$$

for $a = 3$ and all $n \geq 1$, as well as $a = 5$ and 7 for all $n \geq 2$.

The case $a = 3, n \geq 1$. We calculate the Legendre symbol

$$\left(\frac{F_n}{3}\right) = \left(\frac{1 + 2^{2^n} \mod 3}{3}\right) = \left(\frac{1 + (-1)^{2^n}}{3}\right)$$

$$= \left(\frac{2}{3}\right) \equiv 2^{(3-1)/2} \equiv -1 \mod 3$$

From the quadratic reciprocity for the Jacobi symbols we get,—no matter whether F_n is prime or not:

$$\left(\frac{3}{F_n}\right) = \left(\frac{F_n}{3}\right)(-1)^{\frac{(F_n-1)(3-1)}{4}} = \left(\frac{F_n}{3}\right) = -1$$

since $F_n - 1$ is divisible by 4. □

The case $a = 5, n \geq 2$. We calculate the Legendre symbol

$$\left(\frac{F_n}{5}\right) = \left(\frac{1 + 4^{2^{n-1}} \mod 5}{5}\right) = \left(\frac{1 + (-1)^{2^{n-1}}}{5}\right)$$

$$= \left(\frac{2}{5}\right) \equiv 2^{(5-1)/2} \equiv -1 \mod 5$$

□

The case $a = 7, n \geq 2$. The Legendre symbols obey the recurrence

$$\left(\frac{F_n}{7}\right) = \left(\frac{1 + 16^{2^{n-2}} \mod 7}{7}\right) = \left(\frac{1 + 2^{2^{n-2}} \mod 7}{7}\right)$$

$$= \left(\frac{F_{n-2}}{7}\right)$$

for all $n \geq 2$, which allows an induction starting with

$$\left(\frac{17}{7}\right) = \left(\frac{3}{7}\right) \equiv 3^{(7-1)/2} \equiv -1 \mod 7$$

$$\left(\frac{5}{7}\right) \equiv 5^{(7-1)/2} \equiv -1 \mod 7$$

From the quadratic reciprocity for the Jacobi symbols we get,—no matter whether F_n is prime or not:

$$\left(\frac{a}{F_n}\right) = \left(\frac{F_n}{a}\right)(-1)^{\frac{(F_n-1)(a-1)}{4}} = \left(\frac{F_n}{a}\right) = -1$$

since $F_n - 1$ is divisible by 4 and $a = 5, 7$ are odd. □

Proposition 66 (Necessity of a Pépin-like test). *Assume the Fermat number F_n is prime. Then*

$$a^{(F_n-1)/2} \equiv -1 \mod F_n$$

holds for $a = 3$ in the case $n \geq 1$, and $a = 5$ and 7 in the case $n \geq 2$.

Proof. Because of the assumption that F_n is prime, we can use Euler's criterium and get from Euler's criterium

$$a^{\frac{F_n-1}{2}} \equiv \left(\frac{F_n}{a}\right) = -1 \mod F_n$$

as claimed. □

Theorem 23 (Lucas-Lehmer). *Any number $m \geq 2$ is prime if and only if there exists a primarity witness. A witness is a natural number a with the following two properties:*

(i) $a^{m-1} \equiv 1 \mod m$;

(ii) $a^{(m-1)/p} \not\equiv 1 \mod m$ *for all prime divisors p of $m - 1$.*

If $m = F_n$ is a Fermat number, the only prime divisor of $m-1$ is the number 2. Thus we obtain sufficiency for the following Pépin-like test:

Proposition 67 (A more general Pépin-like test). *Assume there exists a natural number a such that*

$$a^{F_n-1} \equiv 1 \mod F_n \quad \text{but} \quad a^{(F_n-1)/2} \not\equiv 1 \mod F_n$$

Then the Fermat number F_n is prime. Moreover, a is a primitive root modulo F_n.

Corollary 34 (A primitive root). *Under the assumptions of proposition 67, the Fermat number F_n is prime. Especially $a = 3$ is a primitive root for the known Fermat primes $5, 17, 257$ and $65\,537$.*

Proof. Let h denote the order of a modulo F_n. This number h is defined to be the minimal number for which holds

$$a^h \equiv 1 \mod F_n$$

By Little Fermat one gets

$$h \mid F_n - 1 = 2^{2^n}$$

This implies $h = 2^t$ for some integer t. Now the assumption

$$a^{(F_n-1)/2} \not\equiv 1 \mod F_n$$

implies $t = 2^n$ and $h = F_n - 1$. Hence a is a primitive root. $\square$

Proof of the Lucas-Lehmer Theorem. The Theorem is true for $m = 2$ because assumption (i) holds for $a = 1$, and assumption (ii) is an empty truth. Assume that a is a witness for $m \geq 3$ and let $h > 1$ be the smallest integer such that $a^h \equiv 1 \mod m$. Assumption (i) yields that h is a divisor of $m - 1$. By assumption (ii), we conclude $h = m - 1$.

Moreover, the witness a is relatively prime to m. The Euler-Fermat Theorem tells that

$$a^{\phi(m)} \equiv 1 \mod m$$

where $\phi(m)$ is the Euler totient function. Hence the order h is a divisor of $\phi(m)$. The inequalities

$$m - 1 = h \leq \phi(m) \leq m - 1$$

imply $\phi(m) = m - 1$. Hence m is a prime number. $\square$

Proth's Theorem is a slight generalization of Pépin's test.

Theorem 24 (Proth's Theorem 1878). *To test whether the number N is prime, one chooses a base a relatively prime to N for which the Jacobi symbol is*

$$\left(\frac{a}{N}\right) = -1$$

Sufficient condition for primality: *Assume additionally that $N = k \cdot 2^m + 1$ with $k \leq 1 + 2^m$. If*

$$(8.0.2) \qquad a^{(N-1)/2} \equiv -1 \mod N$$

then N is prime.

Necessary condition for primarity: *No restriction on k needs to be assumed. If the above congruence (8.0.2) does not hold, then the number N is composite.*

Reason for the sufficient condition: Assume that p is a prime factor of the given number N. Let $h > 1$ be the smallest integer such that $a^h \equiv 1 \mod p$. The assumed congruence (8.0.2) implies that h is a divisor of $N - 1 = k \cdot 2^m$ but not $(N - 1)/2$. The Little Fermat Theorem implies that h is a divisor of $p - 1$. Hence we get the inequalities
$$2^m \leq h \leq p - 1$$
The assumption $k \leq 1 + 2^m$ implies $N \leq 2^{2m} + 2^m + 1 < (1 + 2^m)^2$. Hence
$$\sqrt{N} < 1 + 2^m \leq p \leq N$$
holds for *each prime divisor* of N. Since any composite number has a prime divisor less or equal its square root, this is only possible if N is a prime number. □

The necessary condition. is a direct consequence of Euler's Theorem about the Legendre symbol. □

Here are some further less important remarks about Fermat numbers.

Lemma 84. *For any Fermat number $F_n \neq 3, 5$*
$$2^{(F_n-1)/2} \equiv 1 \mod F_n$$

Proof. By definition
$$2^{2^n} \equiv -1 \mod F_n \quad \text{and} \quad 2^{2 \cdot 2^n} \equiv 1 \mod F_n$$
For all $n \geq 2$ we know that $n + 1 \leq -1 + 2^n$ and hence
$$2^{n+1} = 2 \cdot 2^n \leq 2^{-1+2^n} = (F_n - 1)/2$$
$$2^{2 \cdot 2^n} \equiv 2^{(F_n-1)/2} \equiv 1 \mod F_n$$

□

Lemma 85. *For any Fermat number $F_n \neq 3, 5, 17$*
$$F_{n-1}^{(F_n-1)/2} \equiv 1 \mod F_n$$

Proof. Just calculate:

$$F_{n-1}^2 = (1+2^{2^{n-1}})^2 = 1 + 2 \cdot 2^{2^{n-1}} + 2^{2^n} \equiv 2^{1+2^{n-1}} \mod F_n$$
$$F_{n-1}^{(F_n-1)/2} \equiv 2^{(1+2^{n-1})(F_n-1)/4} \mod F_n$$

Since $n \geq 3$, we know that $n \leq n - 4 + 2^n$ and hence

$$2 \cdot 2^n = 2^{n+1} \leq (1+2^{n-1})2^{-2+2^n} = (1+2^{n-1})(F_n-1)/4$$

hence

$$2^{2 \cdot 2^n} \equiv 1 \mod F_n \quad \text{implies}$$
$$2^{(1+2^{n-1})(F_n-1)/4} \equiv F_{n-1}^{(F_n-1)/2} \equiv 1 \mod F_n$$

□

Lemma 86. *Any Fermat number F_n is a pseudo prime for the base 2 as well as the base F_{n-1}.*

In 1964, Rotkiewicz showed that the product of any number of prime or composite Fermat numbers will be a Fermat pseudo prime to the base 2.

Lemma 87. *Let $n > m \geq 2$. No matter whether the Fermat numbers are prime or not, the Jacobi symbols are*

$$\left(\frac{F_n}{F_m}\right) = \left(\frac{F_m}{F_n}\right) = 1$$

Proof.

$$F_n = 1 + 2^{2^n} = 1 + \left(2^{2^m}\right)^{2^{n-m}} \equiv 1 + (-1)^{2^{n-m}} \equiv 2 \mod F_m$$

for $n > m$. We calculate the Jacobi symbol

$$\left(\frac{F_n}{F_m}\right) = \left(\frac{2}{F_m}\right) = (-1)^{\frac{(F_m-1)(F_m+1)}{8}} = 1$$

since $F_m - 1$ is divisible by 8 for $m \geq 2$. From the quadratic reciprocity for the Jacobi symbols we get,—no matter whether F_n is prime or not:

$$\left(\frac{F_m}{F_n}\right) = \left(\frac{F_n}{F_m}\right)(-1)^{\frac{(F_n-1)(F_m-1)}{4}} = \left(\frac{F_n}{F_m}\right) = 1$$

again since $F_m - 1$ is divisible by 8 for $m \geq 2$. □

Lemma 88. *The number 2 is not a primitive root of any Fermat prime except 3. If $F_n \neq 3, 5, 17$ is a Fermat prime, none of the Fermat numbers $17 \leq F_m < F_n$ with $2 \leq m < n$ is a primitive roots modulo F_n.*

Lemma 89. *Any Fermat prime F_n has $(F_n - 1)/2$ primitive roots. Equivalent are:*

1. *r is a primitive root of F_n.*

2. *r is a quadratic non-residue of F_n.*

3. *$r^{(F_n-1)/2} \equiv -1 \mod F_n$*

The primitive roots of any Fermat prime $F_n \neq 3$ are

$$r \equiv 3 \cdot 9^j \mod F_n$$

with $j = 1, 2, \ldots, (F_n - 1)/2$.

8.0.2 Powers of Three

Problem 108. *Show that the Fermat numbers*

$$F_n = 2^{2^n} + 1 \quad \text{for } n = 0, 1, 2, \ldots$$

are all relatively prime.

Answer. We use $x^2 - 1 = (x+1) \cdot (x-1)$ for $x = 2^{2^k}$ with $k = n-1, n-2 \ldots 0$:

$$F_n - 2 = 2^{2^n} - 1 = (2^{2^{n-1}} + 1) \cdot (2^{2^{n-1}} - 1)$$
$$= (2^{2^{n-1}} + 1) \cdot (2^{2^{n-2}} + 1) \cdot (2^{2^{n-2}} - 1) = \ldots$$
$$\ldots = (2^{2^{n-1}} + 1) \cdot (2^{2^{n-2}} + 1) \cdots (2^{2^0} + 1) \cdot (2^{2^0} - 1)$$
$$= F_{n-1} \cdot F_{n-2} \cdots F_0$$

The common divisor $g := \gcd(F_n, F_k)$ for any $0 \leq k < n$ divides both F_n and $F_n - 2$. Hence $g = 1$ or $g = 2$. But the Fermat numbers are odd, and hence $g = 1$.

Problem 109. *Show that the numbers*

$$D_n = 4^{3^n} + 2^{3^n} + 1 \quad \text{for } n = 0, 1, 2, \ldots$$

are all relatively prime.

Answer. We use $(x^3 - 1)/(x - 1) = x^2 + x + 1$ for $x = 2^{3^k}$ with $k = n - 1, n - 2 \ldots 0$:

$$a - 1 := 2^{3^n} - 1 = \frac{2^{3^n} - 1}{2^{3^{n-1}} - 1} \cdot \frac{2^{3^{n-1}} - 1}{2^{3^{n-2}} - 1} \cdots \frac{2^{3^1} - 1}{2^{3^0} - 1}$$
$$= D_{n-1} \cdot D_{n-2} \cdots D_0$$

The common divisor $g := \gcd(D_n, D_k)$ for any $0 \le k < n$ divides both:

$$g \mid D_n = a^2 + a + 1 \quad \text{and} \quad g \mid a - 1$$

It is easy to see that $\gcd(a^2 + a + 1, a - 1)$ is either 1 or 3. Since $a = 2^{3^n}$ is even, the common divisor cannot be 3. Hence $g = 1$, confirming that the numbers D_k are all relatively prime.

Problem 110. *Let $n \ge 0$. Show that all prime factors of the numbers D_n have the form*

(8.0.3) $$p = 1 + 2 \cdot 3^{n+1} \cdot k$$

Hence all prime factors of D_n and of $2^{3^n} - 1$ are congruent to one modulo six.

Answer. Assume that $p \mid D_n$ is a prime factor of D_n.
Since $D_n \mid N := 2^{3^{n+1}} - 1$ we know that $p \mid N$. Since p is odd, Fermat's Little Theorem says that $p \mid 2^{p-1} - 1$. Hence the little proposition 26 yields

(8.0.4) $$p \mid 2^g - 1 \quad \text{with} \quad g := \gcd(p - 1, 3^{n+1})$$

This implies $g = 3^t$ for some integer $t \le n + 1$. We have even assumed $p \mid D_n$. By Problem 109 the numbers D_k are all relatively prime. Now $p \nmid D_k$ for all $k < n$ implies

$$p \nmid 2^{3^n} - 1 = D_{n-1} \cdot D_{n-2} \cdots D_0$$

Hence $t = n + 1$. Formula (8.0.4) implies

$$\gcd(p - 1, 3^{n+1}) = g = 3^t = 3^{n+1} \quad \text{and hence} \quad 3^{n+1} \mid p - 1$$

and finally relation (8.0.3) is easily confirmed.

Proposition 68. *There exist infinitely many primes $p \equiv 1 \pmod{3}$.*

Try again. We have shown in problem 109 that the numbers
$$D_n = 4^{3^n} + 2^{3^n} + 1 \quad \text{for } n = 0, 1, 2, \ldots$$
are all relatively prime. We have shown in problem 110 that all prime factors of the numbers D_n are congruent to one modulo six. Define the sequence of primes $p_n \mid D_n$ where $p_n > 1$ is the smallest divisor of D_n. These are infinitely many primes $p_n \equiv 1 \pmod{3}$. □

Chapter 9

Prime Testing

9.1 Pseudo Primes, Carmichael Numbers

John Selfridge has conjectured that if p is an odd number, and $p \equiv \pm 2 \pmod 5$, then p will be prime if both of the following hold:

$$2^{p-1} \equiv 1 \pmod p,$$
$$F_{p+1} \equiv 0 \pmod p,$$

where F_k is the k-th Fibonacci number. The first condition is the Fermat primality test using base 2. Selfridge, Carl Pomerance, and Samuel Wagstaff together offer $620 for a counterexample. The problem is still open as of September 11, 2015. [2] Pomerance thinks that Selfridge's estate would probably pay his share of the reward.

Fermat primality test

The simplest probabilistic primality test is the Fermat primality test (actually a compositeness test). It works as follows: Given an integer n, choose some integer a coprime to n and calculate $a^{n-1} \pmod n$. If the result is different from 1, then n is composite. If it is 1, then n may or may not be prime. If $a^{n-1} \equiv 1 \pmod n$ but n is not prime, then n is called a pseudoprime to base a. In practice, we observe that, if $a^{n-1} \equiv 1 \pmod n$, then n is usually prime. But here is a counterexample: if $n = 341$ and $a = 2$, then $2^{340} \equiv 1 \pmod{341}$

even though $341 = 11 \cdot 31$ is composite. In fact, 341 is the smallest pseudoprime base 2 (see Figure 1 of [4]).

There are only 21 853 pseudoprimes base 2 that are less than $2.5 \cdot 10^{10}$ (see page 1005 of [4]). This means that, for n up to $2.5 \cdot 10^{10}$, if $2^{n-1} \equiv 1 \pmod{n}$, then n is prime, unless n is one of these 21 853 pseudoprimes.

Some composite numbers (Carmichael numbers) have the property that $a^{n-1} \equiv 1 \pmod{n}$ for every a that is coprime to n. The smallest example is $n = 561 = 3 \cdot 11 \cdot 17$, for which $a^{560} \equiv 1 \pmod{561}$ for all a coprime to 561. Nevertheless, the Fermat test is often used if a rapid screening of numbers is needed, for instance in the key generation phase of the RSA public key cryptographic algorithm.

Definition 36 (Pseudo prime and Carmichael number). Any number $m \geq 2$ for which

$$(9.1.1) \qquad a^{m-1} \equiv 1 \mod m$$

is called a *pseudo prime* of base $a \geq 2$.

Any number $m \geq 2$ such that equation (9.1.1) holds for all a relatively prime to m is called a *Carmichael number*.

The Carmichael numbers are those which cannot be proved to be composite with the help of the Little Fermat Theorem—if one agrees to use only basis a relatively prime to m.

Proposition 69. *A number m is a pseudo prime with base a if and only if it has the following property:*

(*) *if the prime power p^s divides m, then
p^s divides $a^{\gcd(m-1,p-1)} - 1$.*

Lemma 90. *Especially, if a number m is a pseudo prime with base a and the prime power p^s divides m, then p^s divides $a^{p-1} - 1$.*

Sufficiency of (*). Assume that the base a satisfies the assumption (*).

Let p^s be any prime power dividing m and put $g := \gcd(p-1, m-1)$. Since g divides $m-1$, We conclude by the proposition 26 that $a^g - 1$ divides $a^{m-1} - 1$. The assumption (*) tells that p^s divides $a^g - 1$ and hence $a^{m-1} - 1$ by the the little proposition 26.

The simultaneous congruences $a^{m-1} \equiv 1 \mod p^s$ for all prime power divisors p^s of m together imply $a^{m-1} \equiv 1 \mod m$. □

9.1. PSEUDO PRIMES, CARMICHAEL NUMBERS 245

Necessity of (*). We assume that m is a pseudo prime with base a and thus $a^{m-1} \equiv 1 \mod m$. To check property (*), let p^s be any prime power dividing m. This assumption implies $a^{m-1} - 1 \equiv 0 \mod p^s$, too. The Euler-Fermat Theorem yields $a^{\phi(p^s)} - 1 = a^{p^{s-1}(p-1)} - 1 \equiv 0 \mod p^s$. Hence

$$a^{\gcd(m-1, p^{s-1}(p-1))} \equiv 1 \mod p^s$$

Since p is a divisor of m, it is not a divisor of $m - 1$. Thus $m - 1$ and p^{s-1} are relatively prime and hence

$$\gcd[m - 1, p^{s-1}(p-1)] = \gcd(m-1, p-1)$$

Hence
$$a^{\gcd(m-1, p-1)} \equiv 1 \mod p^s$$

for all prime powers p^s dividing m, as to be shown. □

Lemma 91. *A Carmichael number cannot be divisible neither by any odd prime square nor by 4.*

Proof. Assume that m is a Carmichael number and divisible by the prime power p^s. By the second part of Proposition 91, we know that p^s divides $a^{p-1} - 1$. We now show that it is impossible that any prime square divides m.

Excluding the case $p \geq 3$ is odd and $s \geq 2$: In the case m is divisible by an odd prime square, we choose a to be a primitive root modulo p^2, which we know to exist since p is odd. In that case p^2 is not a divisor of $a^{p-1} - 1$, contradicting that p^s divides $a^{p-1} - 1$.

Excluding the special case $p = 2$ and $s \geq 2$: We need to exclude that m is divisible by 4. We choose $a = 3$. Now 4 is not a divisor of $2 = a^{p-1} - 1$, contradicting that p^s divides $a^{p-1} - 1$.

□

Proposition 70. *Assume that the number m has the property*

(i) *if the prime p divides m, then $p - 1$ divides $m - 1$;*

Assume furthermore that

(a) *the base a is relatively prime to m;*

(b) *if the prime power p^s divides m, then p^s divides $a^{p-1} - 1$.*

Then m is a pseudo-prime of base a.

Proof. Let p^s be any prime power divisor of m. By assumption (b), p^s divides $a^{p-1} - 1$. By assumption (i), $p - 1$ divides $m - 1$ and hence the little proposition 26 implies that $a^{p-1} - 1$ divides $a^{m-1} - 1$. Together we see that p^s divides $a^{m-1} - 1$.

The simultaneous congruences $a^{m-1} \equiv 1 \mod p^s$ for all prime power divisors p^s of m together imply $a^{m-1} \equiv 1 \mod m$. □

Proposition 71. *Assume assumption* (i) *from Proposition 70 holds for the composite number*

$$m = \prod p_i^{s_i} \quad \text{with different primes } p_i.$$

The number of bases a for which the m is a pseudo prime is equal to

$$-1 + \prod (p_i - 1)$$

A power of two is never a pseudo prime.

Proof. For all odd prime factors p, we choose a primitive root r modulo p^s. Choose any integers $0 \leq t < p - 1$, not all zero. By the Chinese Remainder Theorem, the system of simultaneous congruences

$$a \equiv r^{t \cdot p^{s-1}} \mod p^s$$

for all prime factors p has a solution, unique modulo m. It is now easy to check by means of the Euler-Fermat Theorem that a satisfies the assumptions (a) and (b) of Proposition 70. Hence a is a base for the pseudo prime m. The procedure exhausts all possible choices of the bases a. □

Proposition 72. *A number m is a Carmichael number if and only if it has the following two properties:*

(i) *if the prime p divides m, then $p - 1$ divides $m - 1$;*

(ii) *the number m is square free.*

Necessity. We assume that m is a Carmichael number and thus $a^{m-1} \equiv 1 \mod m$ for all a relatively prime to m. To check property (i), let $p \geq 3$ be any odd prime factor of m. We choose a to be a primitive root modulo p. By the little proposition 26

$$\gcd[a^{m-1} - 1, a^{p-1} - 1] = a^{\gcd(m-1,p-1)} - 1$$

9.1. PSEUDO PRIMES, CARMICHAEL NUMBERS

By assumption $a^{m-1} - 1 \equiv 0 \mod p$ and the Little Fermat Theorem yields $a^{p-1} - 1 \equiv 0 \mod p$. Hence

$$a^{\gcd(m-1,p-1)} \equiv 1 \mod p$$

Since a is a primitive root, this implies $\gcd(m-1, p-1) = p-1$. Hence $p-1$ is a divisor of $m-1$, confirming item (i).

The property (ii) has been check above already. □

Sufficiency. Let a be any base relatively prime to m and p be a prime divisor of m. By the Little Fermat Theorem a prime p divides $a^{p-1} - 1$. Since we have assume that $p-1$ divides $m-1$, the little proposition 26 yields that $a^{p-1} - 1$ divides $a^{m-1} - 1$.

Since m is assumed to be square free, the simultaneous congruences $a^{m-1} \equiv 1 \mod p$ for all prime divisors p of m together imply $a^{m-1} \equiv 1 \mod m$, as claimed. □

*Much more has been found out about Carmichael numbers. There exist only finitely Carmichael numbers with three prime factors which can be constructed. The remaining ones have at least four prime factors. Alford et. al. (1994) have proved that the number $C(n)$ of Carmichael numbers less than n has the asymptotics $C(n) \sim n^{2/7}$ for large n.

Proposition 73. *The Carmichael numbers with three prime factors are*

$$(6k + 1)(12k + 1)(18k + 1)$$

were k has to be chosen such that all three factors are primes.

Miller–Rabin primality test

The Miller–Rabin primality test and Solovay–Strassen primality test are more sophisticated variants, which detect all composites (once again, this means: for every composite number n, at least 3/4 (Miller–Rabin) or 1/2 (Solovay–Strassen) of numbers a are witnesses of compositeness of n). These are also compositeness tests.

The Miller–Rabin primality test works as follows: Given an integer n, choose some positive integer $a < n$. Let $2^s d = n - 1$, where d is odd. If

$$a^d \not\equiv 1 \pmod{n} \text{ and}$$
$$a^{2^r d} \not\equiv -1 \pmod{n} \text{ for all } 0 \leq r \leq s - 1,$$

then n is composite and a is a *witness* for the compositeness. Otherwise, n may or may not be prime. The Miller–Rabin test is a strong pseudoprime test.

Solovay–Strassen primality test

The Solovay–Strassen primality test uses another equality: Given an odd number n, choose some integer $a < n$, if

$$a^{(n-1)/2} \not\equiv \left(\frac{a}{n}\right) \pmod{n} \text{ where } \left(\frac{a}{n}\right) \text{ is the Jacobi symbol,}$$

then n is composite and a is a *witness* for the compositeness. Otherwise, n may or may not be prime.

The Solovay–Strassen test is an Euler pseudoprime test. For each individual value of a, the Solovay–Strassen test is weaker than the Miller–Rabin test. For example, if $n = 1905$ and $a = 2$, then the Miller-Rabin test shows that n is composite, but the Solovay–Strassen test does not. This is because 1905 is an Euler pseudoprime base 2 but not a strong pseudoprime base 2. [1]

https://en.wikipedia.org/wiki/AKS_primality_test

---

[1] Daniel J. Bernstein, "Proving Primality After Agrawal-Kayal-Saxena", version of January 25, 2003.

Bibliography

[1] Martin Aigner und Günter M. Ziegler, *Das Buch der Beweise*, Springer, Berlin, Heidelberg, 2002.

[2] David M. Burton, *Elementary Number Theory*, McGraw-Hill, 1997.

[3] Karl F. Gauss, *Disquisitiones Arithmeticae*, Leipzig, 1801.

[4] Robin Hartshorne, *Geometry: Euclid and Beyond*, second printing, Springer, 2002.

[5] Johann Gustav Hermes, *Über die Teilung des Kreises in 65537 gleiche Teile*, Nachrichten von der Gesellschaft der Wissenschaften zu Göttingen, Mathematisch-Physikalische Klasse **3** (1894), 170–186.

[6] Oystein Ore, *Number theory and its history*, McGraw-Hill, 1948.

[7] Magnus Georg Paucker, *Geometrische Verzeichnung des regelmäßigen Siebzehn-Ecks und Zweyhundersiebenundfünfzig-Ecks in den Kreis*, Jahresverhandlungen der Kurländischen Gesellschaft für Literatur und Kunst (1822), 160–219.

[8] Stan Wagon, *Editor's corner: The Euclidean algorithm strikes again*, The American Mathematical Monthly **97** (1990), 125–129.

[9] _____, *A mathematical magic trick*, The College Mathematical Journal **25** (1994), 325–326.

Index

17-gon
    Gauss' formula, 179
17-gon construction, **173–183**
257-gon construction, **206–218**
65 537-gon construction, **218–223**
λ-function, 82

Bertrand's postulate, 45

Carmichael function, 99
Carmichael number, 244, 246
Carmicheal function, 99
ceiling function, 16

division with remainder, 1

Euclidean algorithm, **2**
    convergence speed, 13, 14
    extended, 3
Euclidean property, 19
Euler group, 86
    product of all elements, 151, 153
Euler totient function
    definition, 86
    Euler's Theorem about the, 95
    formula, 87
    lower bound, 88
    recursive solution of $\phi(m) = c$, 89–95
    sum over divisors, 137
    values not assumed, 88

Fermat numbers, **169–239**
    possible prime factors
        Euler 1770, 231
        Lucas 1878, 232
        relatively prime, 231
Fermat primes, 169
    Pépin's test, 233
Fermat's Little Theorem, 84
    consequences of, 85
Fibonacci numbers, 11
field, 102
finite geometric series, 75
floor function, 16

Gauss' easter formula, 16
Gauss' Lemma
    for quadratic reciprocity, 107
Gauss-Wantzel Theorem, 170, 171
greatest common divisor, 1
    of $a^m + 1$ and $a^n + 1$, 78
    of $a^m - 1$ and $a^n - 1$, 76

inclusion-exclusion principle, 26
irrational roots, 23

J-symbol, 119
    algorithm for calculation, 119
    extends the Legendre symbol, 119
    properties implying uniqueness, 121
Jacobi symbols, 123
    extend Legendre symbols, 124
    in mathematica, 125

INDEX 251

recursive calculation, 126
Jacobi symbols for composites
    confirm <u>all</u> residues mod odd
        prime powers, 135
    confirm <u>some</u> non-residues,
        134

Kronecker δ, 82

Lagrange's Theorem
    about polybomials over an
        integral domain, 103
least common multiple, 1
Legendre symbol, 105
    properties, 108
    the basic formula, 106
Legendre's formula
    counting number of primes,
        25

Mersenne prime, 96
Moebius
    function, 81
Moebius inversion
    additive, 81
    multiplicative, 82
    properties, 83
multiplicative function, 87

parity count, 108
    extends the Legendre symbol, 114
partial summation, 42
polynomial ring
    over a field, 102
prim factorization program, 24
prime
    dividing a binomial, 28
prime number, 20
prime numbers
    infinitely many
        another series proof, 37
        Euclid's proof, 20

        Euler's series proof, 40
    infinitely many
        $p \equiv -1 \pmod 4$, 52
    infinitely many
        $p \equiv 1 \pmod 3$, 240
    infinitely many
        $p \equiv 1 \pmod 4$, 231
    infinitely many
        $p \equiv 2 \pmod 3$, 52
    Lucas-Lehmer test, 235
primepower
    dividing a binomial coefficient, 29
    dividing a multinomial coefficient, 30
    dividing $N!$, 28
primitive roots, 135
    criterium for existence, 136
    existence for $2p^r$, 140
    existence for p^2, 141
    existence for composites, 140
    existence for primes, 138
    nonexistence, 142
    number of, 136
product of primes
    lower bound, 49
    upper bound, 35
pseudo prime, 244
Pythagorean triple, **156–167**

quadratic reciprocity
    Gauss' theorem on, 114
quadratic residue
    definition, 104
    Euler's criterium for being
        a, 104
quadratic residues
    modulo composite numbers, 115
    modulo powers of 2, 117

ring, 102

semiprimitive root, 146
simultaneous congruences, 72
Sophie-German prime, 96
square root of complex, 155
Stirling's formula, **35**, **31–35**

uniqueness
    of prime decomposition, 20, 21

Wallis' product formula, 33

www.ingramcontent.com/pod-product-compliance
Lightning Source LLC
Chambersburg PA
CBHW020635220526
45464CB00001B/155